Lecture Notes in Computer Science 8789

Commenced Publication in 1973
Founding and Former Series Editors:
Gerhard Goos, Juris Hartmanis, and Jan van Leeuwen

T0213715

Fernando Bello Stéphane Cotin (Eds.)

Biomedical Simulation

6th International Symposium, ISBMS 2014
Strasbourg, France, October 16-17, 2014
Proceedings

 Springer

Volume Editors

Fernando Bello
Imperial College London, UK
E-mail: f.bello@imperial.ac.uk

Stéphane Cotin
Inria, Strasbourg, France
E-mail: stephane.cotin@inria.fr

ISSN 0302-9743   e-ISSN 1611-3349
ISBN 978-3-319-12056-0   e-ISBN 978-3-319-12057-7
DOI 10.1007/978-3-319-12057-7
Springer Cham Heidelberg New York Dordrecht London

Library of Congress Control Number: Applied for

LNCS Sublibrary: SL 1 – Theoretical Computer Science and General Issues

Typesetting: Camera-ready by author, data conversion by Scientific Publishing Services, Chennai, India

Printed on acid-free paper

Springer is part of Springer Science+Business Media (www.springer.com)

Preface

This book contains the written contributions to the 6th International Symposium on Biomedical Simulation (ISBMS), which was held in Strasbourg, France, during October 16–17, 2014. The 27 articles cover the key scientific areas of this constantly evolving field: soft tissue and fluid modeling, physics-based registration, surgical planning, validation, augmented reality, training systems, and strategies for real-time computation.

Biomedical modeling and simulation are at the center-stage of worldwide efforts to understand and develop computational models of the human body, or at least its principal organs. Large-scale initiatives such as the Physiome Project, Virtual Physiological Human and Blue Brain Project aim to develop advanced computational models that can facilitate the understanding of the behavior of cells, organs, and systems, with the ultimate goal of delivering personalized medicine. At the same time, progress in modeling, numerical techniques, and haptics has enabled more complex and interactive simulations leading to more efficient and safer training of medical personnel. Recently, we are beginning to see the role that real-time computational models can play in the context of computer-assisted interventions, thus bridging the gap between the virtual and real worlds. It is in this context that ISBMS seeks to act as an international forum for researchers where they can share their latest work, discuss future trends, and forge new collaborations.

We received 45 submissions from 11 countries. Each was evaluated by at least three members of the Program Committee. Based on these reviews, 16 manuscripts were selected for long oral presentations and 11 for short talks. The meeting was single track and, in addition to contributed papers, included two keynote presentations, a discussion panel, and live demonstrations of recent research results. The geographical breakdown of the different institutions presenting their research was: Australia, Czech Republic, France, Germany, New Zealand, Romania, South Korea, Spain, Switzerland, UK, and USA. The quality and breadth of the contributions indicates that the symposium continues to be an important forum for our rapidly growing field, bringing together several related disciplines.

We are very grateful to the Program Committee members for volunteering their time to review and discuss the submitted articles and doing so in a timely and professional manner. We are also thankful to the Steering Committee for their encouragement and support in continuing the tradition of a high-quality, focused meeting. We extend our thanks to IHU-Strasbourg, IRCAD, and Inria for providing support in the organization of the meeting. Special thanks go to the local Organizing Committee for their hard work in making the 2014 edition of ISBMS a successful event. Last but not least, we would like to thank all authors for presenting their work at the symposium. It was a pleasure hosting IS-BMS 2014 and we hope that all participants enjoyed the intense and stimulating discussions, as well as the opportunity to establish or renew fruitful interactions.

October 2014   Fernando Bello
                                                                           Stéphane Cotin

Organization

The 2014 edition of ISBMS was organized by the SHACRA team from Inria, the SiMMS group from Imperial College London, and IHU-Strasbourg. It was hosted at IRCAD in Strasbourg, a center of excellence in surgical training.

ISBMS 2014 would not have been possible without the dedication and hard work of the Organizing Committee: Hadrien Courtecuisse (CNRS), Alejandro Granados (ICL), Rosalie Plantefève (Inria), and Audrey Ziliox (IHU).

General Chairs

Fernando Bello	Imperial College London, UK
Stéphane Cotin	Inria, France
Jérémie Dequidt	University of Lille, France
Igor Peterlik	IHU, France and Masaryk University, Czech Republic

Steering Committee

Stéphane Cotin	Inria, France (Chair)
Fernando Bello	ICL, UK (Co-chair)
Nicholas Ayache	Inria, France
Steve Dawson	CIMIT, USA
Hervé Delingette	Inria, France
Matthias Harders	University of Innsbruck, Austria
Dimitris Metaxas	Rutgers University, USA
Gabor Szekely	ETH, Switzerland

Program Committee

Cagatay Basdogan	Koc University, Istanbul, Turkey
Sofia Bayona	University Rey Juan Carlos, Madrid, Spain
Stéphane Bordas	University of Luxembourg, Luxembourg
Hadrien Courtecuisse	Icube Laboratory, CNRS, Strasbourg, France
Elias Cueto	University of Zaragoza, Spain
Hervé Delingette	Inria, Sophia Antipolis, France
Jérémie Dequidt	University of Lille, France
Christian Duriez	Inria, Lille, France
Philip 'Eddie' Edwards	Imperial College London, UK
Derek Gould	University of Liverpool, UK

Table of Contents

Training Systems and Haptics

Physics-Based Registration

Vascular Modelling and Simulation

Image and Simulation

Surgical Planning

Modelling

Analysis, Characterisation and Validation

Preliminary Bone Sawing Model for a Virtual Reality-Based Training Simulator of Bilateral Sagittal Split Osteotomy

Thomas C. Knott[1], Raluca E. Sofronia[2], Marcus Gerressen[3], Yuen Law[1],
Arjana Davidescu[2], George G. Savii[2], Karls H. Gatzweiler[4],
Manfred Staat[4], and Torsten W. Kuhlen[1]

[1] Virtual Reality Group, RWTH Aachen University, Germany
[2] Department of Mechatronics, Politehnica University of Timisoara, Romania
[3] Department of Oral Maxillofacial and Plastic Facial Surgery,
University Hospital of Aachen, Germany
[4] Biomechanics Laboratory, Aachen University of Applied Sciences, Germany
knott@vr.rwth-aachen.de
http://www.vr.rwth-aachen.de

Abstract. Successful bone sawing requires a high level of skill and experience, which could be gained by the use of Virtual Reality-based simulators. A key aspect of these medical simulators is realistic force feedback. The aim of this paper is to model the bone sawing process in order to develop a valid training simulator for the bilateral sagittal split osteotomy, the most often applied corrective surgery in case of a malposition of the mandible. Bone samples from a human cadaveric mandible were tested using a designed experimental system. Image processing and statistical analysis were used for the selection of four models for the bone sawing process. The results revealed a polynomial dependency between the material removal rate and the applied force. Differences between the three segments of the osteotomy line and between the cortical and cancellous bone were highlighted.

Keywords: Bone sawing, virtual reality, training simulator, bilateral sagittal split osteotomy.

1 Introduction

Bone sawing is used in many medical procedures, such as: ostectomies, osteotomies, harvesting of bone grafts, arthroplasties and amputations. The success of the surgery and the rate of recovery of the patients are closely related to the precision of the sawing process. The complex structure and the anisotropy of the bone [1] can affect the sawing process. Concerning these, a high level of dexterity is required on part of the surgeon. During sawing procedure, they therefore rely a lot on tactile and force feedback, which leads to the demand of high-fidelity realistic haptic feedback in the training systems. Regardless of the

F. Bello and S. Cotin (Eds.): ISBMS 2014, LNCS 8789, pp. 1–10, 2014.

traditional training methods, computer-based simulations have proven to be a valid alternative. In case of Virtual Reality (VR)-based training systems, a virtual environment is created by a computer and the trainee is able to naturally interact with it, e.g. by physical interaction by means of a haptic interface.

For simulation, bone cutting is often related with metal machining due to the strong similarity between the tools used [2, 3]. Besides this, other physics-based approaches were used for cutting force models in VR-based training systems, such as: Hertz's contact theory [4] or impulse theorem and Coulomb's law of dry friction [5]. Several authors have studied the specific cutting energy of the bone in certain processes [6, 7], but no definitive conclusion can be formulated at the present time in the case of bone sawing. Other authors used specific bone properties, e.g. bone mineral density [8], to establish a relation for the drilling forces. While there are numerous studies investigating the bone mechanical properties, there are only a few studies conducted on particular case, such as the human jaw [9–13], due to the difficulty of obtaining bone test samples with the dimensions required by most testing standards.

Concerning the treatment of deformities or malposition of human jaws, unfavourable sawing pattern or insufficient area of contact can lead to procedure-related errors, which can range from aesthetically unpleasant results to malocclusion and even life-threatening bleedings. Among the maxillofacial surgery procedures, bilateral sagittal split osteotomy (BSSO), according to Obwegeser and Dal Pont, is probably the most frequently used technique for total osteotomy of the mandible [14]. It is performed via an intraoral approach and starts with detaching the soft tissue from the mandibular ramus and body. After that, the osteotomy line is marked using a saw or a Lindemann's burr and successively deepened. By reversed twisting of two chisels inserted into the line, the mandible is split apart. After relocation in the desired position, the segments are fixed using an appropriate osteosynthesis. The technical difficulty of the procedure requires a high level of dexterity and experience which can be gained only through training. Currently, the training of the BSSO is limited to human cadavers or to patients. The first alternative is expensive and not readily available; the second alternative entails obvious risks for the patients. Due to these considerations, the aim of this study is to obtain a physics-based model for bone sawing during BSSO in order to implement it into a VR-based training system for the mentioned procedure [17].

The paper is organized in five sections. The next section, Materials and Methods, contains an analysis of a typical sawing process for maxillofacial surgery, in order to determine the process parameters. Furthermore, the conducted experiments, made to gain realistic values for the parameters, and the according methods used for data processing and statistical analysis are described as well. Afterwards, the obtained regression models for bone sawing are presented and discussed.

2   Materials and Methods

2.1   Bone Sawing Analysis

All material machining processes (turning, milling, drilling, sawing, etc.) can be reduced to the general case of oblique cutting with a single point cutting tool. The tool at a rake angle of α is moved against the workpiece with a velocity v in order to remove a layer of material in the form of a chip. The depth of the layer removed by the tool is known as the undeformed chip thickness s (see Fig. 1).

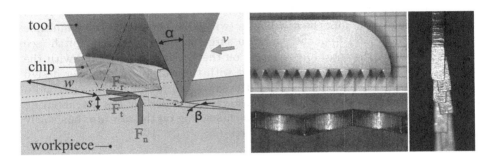

Fig. 1. (Left) Components of the cutting force. (Right) Sawing Blade.

In the general case of oblique cutting (helix angle β different than zero), the force that acts on the tool has three components [15, 16]: the tangential force F_t (in the cutting direction), the normal force F_n (in the feeding direction) and the radial force F_r (perpendicular on both forces F_n and F_t) (see Fig. 1). These forces are proportional to the section area of the undeformed chip A (the product of the chip thickness s times the chip width w [2, 15]:

$$\begin{bmatrix} F_t \\ F_n \\ F_r \end{bmatrix} = \begin{bmatrix} K_t \\ K_n \\ K_r \end{bmatrix} \cdot A \tag{1}$$

where K_t, K_n and K_r $[Nm/m^3]$ are the specific cutting energies along the tangential, normal and radial direction of the cutting. The specific cutting energy can vary considerably for a given material due to the influence of the cutting conditions, e.g. rake angle, helix angle, cutting velocity etc. However, for small changes in the cutting conditions, at a high cutting velocity and large feeds, the specific cutting energy tends to be constant and can be used like a mechanical property of the material [15].

In case of reciprocating sawing, the typically cutting process in maxillofacial surgery, the material removal process is the effect of two motions: (i) an oscillating motion along the workpiece, and (ii) a linear motion, the feed, moving the saw into the workpiece. The process is characterized by small feeds, high sawing speeds and kerf formation. Therefore, the sawing force for one tooth can be better related to the volume of the undeformed chip V_i. The volume can be determined by analysing the saw-tooth movements. The tooth moves in two successive positions during a period of time equal to half of the reciprocating frequency f, along the workpiece (with a distance equal to the reciprocating amplitude, the stroke, a) and along the feeding direction (with a distance equal to the chip thickness s). Therefore, the volume of the undeformed chip is:

$$V_i = w \cdot a \cdot s = w \cdot a \cdot \frac{v}{2 \cdot f} \tag{2}$$

where w is the chip width (equal to the saw blade kerf) and v is the feeding speed. According to the cutting tool classification [16], the saw is considered a multiple cutting edge tool. The saw blades (which are considered to be multiple cutting edge tools [16]) frequently used in surgeries have a constant pitch (see Fig. 1). Therefore, the total forces acting on the saw are proportional with the number of saw-teeth n that are in contact with the workpiece. Another design aspect of the surgical saw blade is that the saw teeth are facing opposite sides (see Fig. 1) in order to cancel the radial forces generated by the neighbouring teeth. Therefore, the cutting force components acting on the saw are then solely the tangential force F_t and the normal force F_n:

$$\begin{bmatrix} F_t \\ F_n \end{bmatrix} = \begin{bmatrix} K_t \\ K_n \end{bmatrix} \cdot V = \begin{bmatrix} K_t \\ K_n \end{bmatrix} \cdot n \cdot V_i \tag{3}$$

where V is the volume removed by the saw, K_t and K_n are the tangential and normal force parameters, respectively. The number of saw-teeth that act on the workpiece can be obtained by dividing the length of the saw part l that is in contact with the workpiece to the saw pitch p. The product of the length of the cut, the incision saw blade width and the cutting speed, is called the removal rate $R_r = l \cdot w \cdot v$. According to these considerations the forces acting on the saw are:

$$\begin{bmatrix} F_t \\ F_n \end{bmatrix} = \begin{bmatrix} K_t \\ K_n \end{bmatrix} \cdot \frac{l}{p} \cdot w \cdot a \cdot \frac{v}{2 \cdot f} = \frac{1}{2} \cdot \begin{bmatrix} K_t \\ K_n \end{bmatrix} \cdot \frac{R_r \cdot a}{p \cdot f} \tag{4}$$

Due to the reciprocating saw design, the tangential force is reduced, in the user hand, to a high-frequency vibration. In conclusion, the force perceived by the user for a small reciprocating saw typically used in maxillofacial surgery depends on the material removal rate and the saw parameters (saw pitch, frequency and amplitude of the saw reciprocating motion):

$$F_n = K_n \cdot \frac{R_r \cdot a}{p \cdot f} \tag{5}$$

2.2   Testing System

In order to establish a quantitative relation for bone sawing, a testing system was designed (see Fig. 2). A bone sample fixed on the carriage of a guidance system was pulled into the saw by the use of a pulley system. The forces exerted on the bone sample were acquired using a K3D120 three-dimensional force sensor from ME-Messsysteme GmbH (range 50 N, resolution 0.01 N, eigenfrequency 1000 Hz). The relative distance between the bone sample and the saw was measured by the PMI80-F90-IU-V1 inductive distance sensor from Pepperl and Fuchs (range 80mm, resolution 125 µm). The measurements were collected and stored at a frequency of 500 Hz using the GSV-3USBx2 2mV/V data acquisition system from ME-Messsysteme GmbH. A program for data acquisition was developed using LabVIEWiew 8.5.1 from National Instruments. The sawing system used in the experiments was the Bran Aesculap Microspeed Artho system composed of: a GD678 motor, a GB130R small reciprocating saw (frequency 20000 rpm, stroke 3 mm) and a GC909R saw blade (cutting length 33 mm, width 0.4 mm and kerf 0.6 mm).

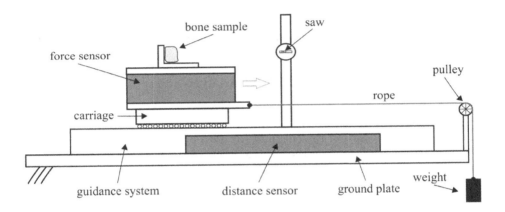

Fig. 2. The testing system

2.3 Bone Sample Preparation

A human cadaver mandible (male cadaver, 58 years old) was used for the experiments. The mandible was sectioned in six samples: two samples for each of the three typical osteotomy line segments (see Fig. 3). The bone samples were fixed with glue and screws on the force sensor through a mechanical connector, in order to obtain the same cutting directions such as the ones of the BSSO (see Fig. 3).

2.4 Testing Conditions

Due to the fact that only one mandible was used for testing, we used the same reciprocating motion parameters (constant reciprocating frequency and amplitude) and the same saw blade (constant saw pitch). In order to select the force for pulling the bone into the saw, we determined typical sawing forces in pre-measurements with an expert surgeon. For each of the bone samples a number

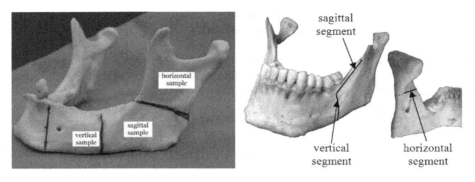

Fig. 3. Bone sample preparation according to the osteotomy line. (Left) The location of the samples. (Right) The segments of the osteotomy line.

of cuts (2...3 cuts) were made according to the procedure specifications and the available space.

2.5   Data Processing and Statistical Analysis

To get the final parameters the collected data was processed in two steps: (i) image processing of the section cuts in order to obtain the dependency between the depth and the length of the cut, and (ii) data processing to relate the measured normal sawing forces via the sawing depth, cut length, and feed rate to a removal rate. Up to three areas were determined for each section cut according to the structure of the bone: cortical bone area, cancellous bone area and mixed cortical-cancellous bone area. Only the areas where one type of bone exists were used in the analysis. Due to the particularities of the osteotomy line, four cases were analysed: cortical bone in the horizontal cut (c-h), cortical bone in the vertical cut (c-v), cortical bone in the sagittal cut (c-s) and cancellous bone in the sagittal cut (s-s).

For each of the four cases, the regression models of the dependency between the removal rate and the normal force were determined. The goodness of the fits was analysed based on the coefficient of determination R^2. Furthermore, the non-linear correlation coefficient (measure of the strength of a non-linear relation between two variables) was computed for each case in order to illustrate the need of a non-linear fit for the removal rate versus force dependency.

3   Results

The dependencies between the removal rate and the normal sawing force are shown in Figure 4. A non-linear dependency can be observed directly from the graphics in each of the four cases. The non-linear correlation coefficients were: (c-h) 0.981; (c-v) 0.984; (c-s) 0.961; (s-s) 0.975.

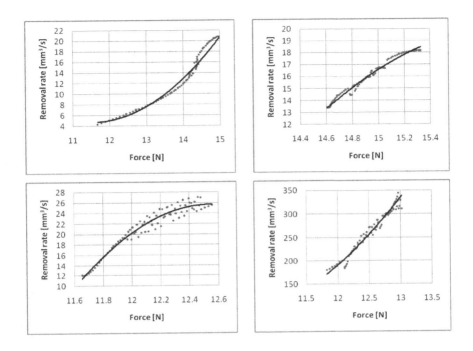

Fig. 4. The bone removal rate versus the force. Legend: (•) the experimental points; (-) the regression model. (Top left) Cortical bone from the horizontal cut. (Top right) Cortical bone from the vertical cut. (Bottom left) Cortical bone from the sagittal cut. (Bottom right) Cancellous bone from the sagittal cut.

Based on the goodness of the fit, the following regression models of the removal rate $R_r[\text{mm}^3/\text{s}]$ versus force $F_n[\text{N}]$ were chosen to represent the dependencies (6,7,8,9): (c-h) Cortical bone from the horizontal cut ($R^2 = 0.969$):

$$R_r = 1.263 \cdot F_n^2 - 28.859 \cdot F_n + 169.43 \tag{6}$$

(c-v) Cortical bone from the vertical cut($R^2 = 0.978$):

$$R_r = -3.614 \cdot F_n^2 + 115.21 \cdot F_n - 898.47 \tag{7}$$

(c-s) Cortical bone from the sagittal cut ($R^2 = 0.941$):

$$R_r = -17.548 \cdot F_n^2 + 440.72 \cdot F_n - 2741.6 \tag{8}$$

(s-s) cancellous bone from the sagittal cut ($R^2 = 0.954$):

$$R_r = 34.063 \cdot F_n^2 - 706.19 \cdot F_n + 3762.4 \tag{9}$$

The regressions were significantly different between the cortical and the cancellous bone; the values of the cancellous removal rate were much higher than the ones of the cortical bone removal rate. A difference in the regressions for the three segments of the osteotomy line was present, e.g. for the same removal rate of 15...20 mm^3/s the required force was higher for the vertical cut than for the horizontal cut, which was also higher than the force required for the sagittal cut.

4   Discussion

In this study, the cutting process with a small reciprocating saw was analysed in order to establish the process parameters. The analysis results showed that the cutting force depends on the removal rate and on the cutting parameters (saw pitch, frequency and amplitude of the saw reciprocating motion). Experiments conducted on the particular case of the BSSO line sawing proved the dependency between the removal rate and the force.

The removal rate of the cancellous bone was found considerably higher than the one of the cortical bone. This is the consequence of the significant differences between the mechanical properties of these bone structures [9, 11, 12]. Furthermore, variations between the three osteotomy lines were observed. The variations can be attributed to the location sites within the mandible. Schwartz-Dabney et al. [9] reported slightly smaller values for the mechanical properties in the region and on the direction of the horizontal cut than in the one of the vertical cut. Additionally, within a region, the properties vary due to the anisotropy of the bone [9]. It is reported that the mandible is stiffer and stronger in the longitudinal direction than in the radial and tangential direction, mainly due to the orientation of the osteons which are primarily oriented in his direction [12, 13]. However, as the determined removal rates are specific for the cutting conditions used in the experiments, which are dictated by the BSSO use case, it is not possible to directly compare these values to generic mechanical properties of the mandible reported in the literature.

The differences between the theoretical model and the experimentally gained regression models can be attributed to the (i) limitations of the theoretical model (constant specific energy for small changes in the cutting conditions at a high cutting velocity and large feeds [15]), (ii) the heterogeneous properties of the bone, (iii) the smaller feeds and (iv) the chip formation mechanism due to the saw blade kerf. Although, the non-linear model is not based on the theoretical model it is useful to represent an essence of the experimentally gained data in a compact way for the usage in the sawing simulation (see also below).

The limitations of this study are related to the number of sample cuts and mandibles. Due to the fact that only one mandible was used for testing in this stage of the research, only a narrow number of configurations could be taken into consideration (limited force variance and constant sawing parameters). The future work will focus on extending the models by testing a statistically meaningful number of cases and by taking into consideration the neglected parameters. Based on these measurements a more generic model could then be developed which may also consider factors besides sawing conditions like patient age etc.

4.1 Usage in a Training Simulator

The presented results can be utilized in a virtual reality-based simulator for training of the BSSO with a specific bone scenario. In [17] a simulator prototype is described where a trainee can interact via a haptic input device with a voxel-based sawing simulation. In this, a material removal component takes off bone material depending on the interaction forces between the saw and the bone as well as their collision configuration (for more details, we refer the interested reader to [17]). The configuration specific measurements relating normal forces and bone removal rates presented above, can be integrated into this component to create a realistic sawing behavior. The according removal rate for an interaction force can thereby be determined by searching for a set of appropriate close by measurements and calculating a weighted average. An computationally less expensive alternative is the utilization of the closed-form formulas given by the non-linear models.

5 Conclusion

To enable the development of realistic VR-based simulators which could improve surgeons bone sawing skills a good model of the underlying process is required. This research presents several mathematical models to quantify the material removal rate according to the force applied to the saw in contact with the bone. The proposed models refer to the particular case of a maxillofacial surgery procedure, the BSSO. The presented data reveals differences between the three osteotomy lines and the cortical and cancellous bone, which is consistent with previous findings about the mechanical properties of the human mandible.

Acknowledgments. This work was funded by the German Research Foundation under grants: KU 1132/6-1, FR 2563/2-1 and GE 2083/1-1. Additionally, it was partially supported by the strategic grant POSDRU/88/1.5 /S/50783, Project ID50783 (2009), co-financed by the European Social Fund- Investing in People 2007-2013. Furthermore, it received funding from the European Unions Seventh Framework Programme for research, technological development and demonstration under grant agreement no 610425.

References

1. Brydone, A.S., Meek, D., Maclaine, S.: Bone grafting, orthopaedic biomaterials, and the clinical need for bone engineering. Proceedings of Inst. Mech. Eng. H 224, 1329–1343 (2010)
2. Arbabtafti, M., Moghaddam, M., Nahvi, A., et al.: Physics-based haptic simulation of bone machining. IEEE Transactions on Haptics 4(1), 39–50 (2011)
3. Hsieh, M.S., Tsai, M.D., Yeh, Y.: An amputation simulator with bone sawing haptic interaction. Biomedical Engineering: Applications, Basis and Communications 18, 229–236 (2006)

4. Agus, M.: Haptic and visual simulation of bone dissection. Ph.D. Thesis, Universita degli Studi di Cagliari, Italy (2004)
5. Wang, Q., Chen, H., Wu, J.H., et al.: Dynamic touch-enable bone drilling interaction. In: 5th International Conference on Information Technology and Applications in Biomedicine, Shenzhen, China, pp. 457–460 (2008)
6. Jacobs, C.H., Pope, M.H., Berry, J.T., et al.: A study of the bone machining process orthogonal cutting. Journal of Biomechanics 7(2), 131–136 (1974)
7. Plaskos, C., Hodgson, A.J., Cinquin, P.: Modelling and optimization of bone-cutting forces in orthopaedic surgery. In: Ellis, R.E., Peters, T.M. (eds.) MICCAI 2003. LNCS, vol. 2878, pp. 254–261. Springer, Heidelberg (2003)
8. Ong, F.R., Bouazza-Marouf, K.: Evaluation of bone strength: Correlation between measurements of bone mineral density and drilling force. Proceedings of the Institution of Mechanical Engineers. Part H, Journal of Engineering in Medicine 214(4), 385–399 (2000)
9. Schwartz-Dabney, C.L., Dechow, P.C.: Variations in cortical material properties throughout the human dentate mandible. American Journal of Physical Anthropology 120, 252–277 (2003)
10. Seong, W.J., Kim, U.K., Swift, J.: Elastic properties and apparent density of human edentulous maxilla and mandible. International Journal of Oral and Maxillofacial Surgery 38, 1088–1093 (2009)
11. Misch, C.E., Qu, Z., Bidez, M.W.: Mechanical properties of trabecular bone in the human mandible: implications for dental implant treatment planning and surgical placement. Journal of Oral and Maxillofacial Surgery 57(6), 700–706 (1999)
12. Van Eijden, T.M.: Biomechanics of the mandible. Critical Reviews in Oral Biology and Medicine 11, 123–136 (2000)
13. Nomura, T., Gold, E., Powers, M.P., et al.: Micromechanics/structure relationships in the human mandible. Dental Materials 19, 167–173 (2003)
14. Gerressen, M., Zadeh, M.D., Stockbrink, G., et al.: The functional long-term results after bilateral sagittal split osteotomy (BSSO) with and without a condylar positioning device. Journal of Oral and Maxillofacial Surgery 64, 1624–1630 (2006)
15. Boothroyd, G., Knight, W.A.: Fundamentals of Machining and Machine Tools. Marcel Dekker, New York (1989)
16. Groover, M.P.: Fundamentals of Modern Manufacturing: Materials, Processes and Systems, 4th edn. Wiley, New York (2010)
17. Sofronia, R., Knott, T., Davidescu, A., Savii, G., Kuhlen, T., Gerressen, M.: Failure mode and effects analysis in designing a virtual reality-based training simulator for bilateral sagittal split osteotomy, International Journal of Medical Robotics and Computer Assisted Surgery 9(1), e1–9 (2013)

Interactive Training System
for Interventional Electrocardiology Procedures

Hugo Talbot[1,2], Federico Spadoni[2], Christian Duriez[1], Maxime Sermesant[2],
Stephane Cotin[1], and Hervé Delingette[2]

[1] Shacra Team, Inria Lille, North Europe, France
[2] Asclepios Team, Inria Sophia Antipolis, Méditerranée, France

Abstract. Recent progress in cardiac catheterization and devices allowed to develop new therapies for severe cardiac diseases like arrhythmias and heart failure. The skills required for such interventions are still very challenging to learn, and typically acquired over several years. Virtual reality simulators can reduce this burden by allowing to practice such procedures without consequences on patients. In this paper, we propose the first training system dedicated to cardiac electrophysiology, including pacing and ablation procedures. Our framework involves an efficient GPU-based electrophysiological model. Thanks to an innovative multithreading approach, we reach high computational performances that allow to account for user interactions in real-time. Based on a scenario of cardiac arrhythmia, we demonstrate the ability of the user-guided simulator to navigate inside vessels and cardiac cavities with a catheter and to reproduce an ablation procedure involving: extra-cellular potential measurements, endocardial surface reconstruction, electrophysiology mapping, radio-frequency (RF) ablation, as well as electrical stimulation. This works is a step towards computerized medical learning curriculum.

1 Introduction

Among all cardiovascular diseases, cardiac arrhythmia and heart failure are life-threatening pathologies. Cardiac arrhythmia consists in an abnormal electrical activity in the myocardium (heart walls). Depending on the pathology, different therapies are pursued. RF ablation is preferred for ventricular tachycardia. In this paper, we only consider ventricular extrasystole, i.e. ventricular tachycardia caused by ectopic foci. An ectopic focus is an abnormal pacemaker area (outside of the sinoatrial node) that initiates abnormal self-generated beats. Such pathologies can occur upon changes in the heart structure following a coronary artery disease or as chronic consequences of hypertension, diabetes or cardiomyopathy, as stated in [9].

Considering an ectopic focus located inside the right ventricle (RV), the procedure first consists in inserting catheters from the femoral vein up to the RV under fluoroscopic imaging. Once inside the ventricle, an electrophysiology mapping is performed by exploring the endocardial surface with catheters to map the activation patterns. These patterns allow to locate the ectopic focus responsible for the arrhythmia. Each pathological region found by electrophysiology

F. Bello and S. Cotin (Eds.): ISBMS 2014, LNCS 8789, pp. 11–19, 2014.

mapping will eventually be ablated using RF: heating the cardiac tissue next to the ectopic focus leads to cellular death, thus suppressing the related abnormal beats.

Until now, residents in cardiology train on patients by separately learning each step of the procedure under the supervision of a senior cardiologist. In order to shorten the training period and to allow a virtual training on complex patient cases, we propose a training system for interventional cardiology based on the simulation of electrophysiology.

Previous Work. Previous research projects already led to training simulators in cardiology as for instance detailed in [4]. The most recent simulator detailed in [3] focuses on intra-ventricular navigation. Main contributions of [3] include the reproduction of the conditions for the slip and nonslip interaction of the catheter. Authors present a qualitative analysis of the catheterization training using experimental data on a porcine left ventricle, as well as an user evaluation. Few training systems succeeded to be commercialized such as Cathi from Siemens, VIST (Vascular Intervention Simulation Trainer) from Mentice, Simantha from SimSuite, CathLabVR from CAE HealthCare and Angio Mentor from Simbionix. All these simulators are mostly focusing on the endovascular navigation including pre-recorded electrocardiograms (ECG), but none of these simulators neither includes a biophysical modeling of the cardiac electrophysiology, nor the interaction with the electrophysiology model.

Simulating the human cardiac electrophysiology is a wide field of research. However, only recent work [11, 12] investigate high performance computing applied to cardiac electrophysiology to achieve quasi real-time simulations. Therefore, coupling a simulation of endovascular navigation with a model of cardiac electrophysiology while keeping performances close to real-time is extremely challenging.

Based on the Mentice VIST hardware (see Fig. 1(a)), we present a training system dedicated to interventional electrocardiology procedures that combines a biophysical modeling of cardiac electrophysiology with an endovascular catheterization. Based on our GPU model of cardiac electrophysiology, the training framework simulates the electro-anatomical mapping, perform RF ablation and electrical stimulation in real-time. Finally, a performance analysis is conducted based on a synthetic case of ventricular extrasystole.

2   Material and Methods

2.1   4D-Image Based Model of the Heart

The simulator is developed using SOFA[1] and reproduces an intervention aiming at ablating an ectopic focus located in the RV. For a realistic navigation, the

---

[1] SOFA is an open source framework for interactive numerical simulations in medicine. More information about SOFA can be found at http://www.sofa-framework.org

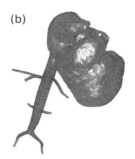

Fig. 1. (a) Our setup using the Mentice VIST device; (b) Resulting triangular mesh used cardiovascular navigation

cardiovascular anatomy is needed.Patient data were acquired in the framework of the European euHeart project[2]. 3D MRI are preoperatively obtained in order to reconstruct the patient-specific heart anatomy. Both atria and ventricles are labeled using a plugin tool implemented in GIMIAS[3]. The mask resulting from the segmented steady-state free precession magnetic resonance images (MRI) is then meshed using the CGAL library[4]. Dedicated to endovascular navigation, a first mesh generation of the four cardiac chambers is built including 12,950 triangular elements.A second, static and finer mesh, only modeling the ventricles, is extracted for the electrophysiology computation including 30,807 linear tetrahedra.

However, the catheter navigation simulation also requires a mesh of the venous system leading to the heart, since navigation starts from the femoral vein. Regarding blood vessels, synthetic data of the inferior vena cava are extracted from the Zygote data set[5]. The resulting mesh for navigation fuses the generic model of vena cava with the patient-specific mesh of the heart (presented in Fig. 1(b)).

After re-ordering the cardiac phases (from passive filling to ventricular isovolumetric relaxation), we estimate the cardiac motion from 4D cine MRI information using a Demon-based registration algorithm detailed in [8]. The estimated deformation field is resampled on the vertices of the navigation mesh, thus resulting in a realistic beating heart model.

2.2   Catheter Navigation Model

The real-time simulation of the catheter behavior during endovascular procedures is particularly challenging and has been the central interest of several

---

[2] For more information about the euHeart project: www.euheart.eu

[3] GIMIAS is an open source framework providing image visualization, manipulation, and annotation. For more information: www.gimias.net

[4] CGAL is an open source software library that provides algorithms in computational geometry. For more information: www.cgal.org

[5] The Zygote data are a set of 3D anatomical models sold by the company Zygote Media Group. For more information: www.zygote.com

research work. Catheters are wire-like structures characterized by stiff and light materials, high tensile strength and low resistance to bending. Our work relies on the FEM introduced by Duriez et al. in [6] and further developed by Dequidt et al. in [5], both in the context of coil embolization in neurology. This method is based on Kirchhoff rod theory, so that our catheter results in a serial set of beam elements. We choose this corotational approach since it handles geometric non-linearity due to large changes in the shape of the object. Other models exist for wire-like structures as the inextensible super-helices model proposed by Bertails et al. [2] or a linear representation of angular springs in Wang et al. [13]. However, super-helices from Bertails involve a quadratic time complexity regarding the number of helical elements. In comparison, corotational beam approach offers a linear complexity. Based on non-physical angular springs, the representation chosen by Wang implies non-physical behavior. The corotational model assumes that the strains remain "small" in a local frame defined at the level of each element. In our scope of endovascular navigation, the catheter undergoes large displacements but only small strains, which meets the corotational assumption.

The collision detection is performed using first a bounding volume hierarchy (BVH) for the broad phase, and then computing local minimal distances, as introduced in [7]. Based on the Signorini's law combined with the Coulomb's law, the constraint resolution leads to a non-linear complementarity problem solved using a Gauss-Seidel algorithm.

2.3   GPU Electrophysiology Model

Our work is based on the Mitchell Schaeffer (MS) [10] model since (i) it has only 5 parameters, (ii) each parameter has a physiological meaning and (iii) it provides a better estimation of the action potential compared to other phenomenological models (as the Aliev-Panfilov model [1]). Since it only captures the transmembrane potential V_m, the MS model is a "mono-domain" model. Only "bi-domain" models can simulate both intra-cellular U_i and extra-cellular potentials U_e, where $V_m = U_i - U_e$.

To increase computation efficiency, the electrical activity of the heart is only simulated for the ventricles. This consideration is acceptable since atria and ventricles are electrically isolated by a collagen layer. In our model, the stimulation is induced by the Purkinje fibers. Relying on the finite element method, the ventricular electrophysiology is computed on a static mesh using 30,807 elements. Moreover, the implementation of the weak form of reaction diffusion equations leads to zero Neumann boundary conditions, i.e. the electrical current is null in the orthogonal direction of the border. Based on [12], the entire electrophysiology model is implemented on GPU. The simulation time step is constrained by the coarseness of the mesh. In this simulation, we use a full explicit backward differentiation integration scheme with a time step $dt_E = 10^{-4}$ s.

To integrate both electrophysiology and navigation simulations, we exploit the CPU parallelism and we choose a task scheduling architecture. This technique is an efficient way to scale the computation to all the CPU cores available. The main loop of our multithreading architecture is split into a parallel part executing concurrently the endovascular navigation and electrophysiology simulations and a serial part executing the graphics rendering. Electrophysiology simulation requires a very low time step $dt_E \leq 1.5 \cdot 10^{-4}$ s for stability reasons, whereas navigation simulation is running with $dt_N \leq 0.02$ s. Our framework can then be depicted as a asynchronous simulation based on multithreading.

2.4   Interactive Model

In the scope of ventricular extrasystole, the most time consuming step of the RF ablation procedure consists in locating and ablating the ectopic focus. This area includes excitable cells initiating a premature heart beat, or ectopic beat. Once catheters are positioned inside the heart cavities, the cardiologist needs to reconstruct an electro-anatomical mapping by acquiring activation times on the endocardium. The pathological area can be localized since the ectopic focus corresponds to the region with the earliest activation time. The complexity of this procedure highlights the importance of interacting with the electrophysiology *in silico*.

Extra-Cellular Potential Measurements. During the procedure, cardiologists use catheters to interpret the electrical activity of the heart. These catheters can measure either unipolar or bipolar potentials, respectively measuring the extra-cellular potential U_e or a difference of extra-cellular potentials ΔU_e. Bipolar potentials are often preferred as they remove the far field potentials and provide sharper depolarizations. Since the MS model is mono-domain, the four-electrode catheter only records transmembrane potential (see Fig. 3(a)).

A realistic modeling of the displayed signals is key for our training simulator since cardiologists mainly rely on electrophysiological signals to understand the

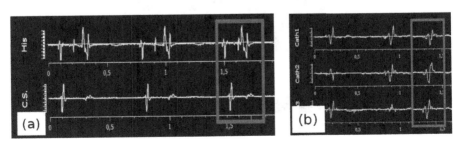

Fig. 2. (a) Unipolar measured at the His bundle and coronary sinus; (b) Bipolar signals recorded by the four electrodes

cardiac arrhythmia and guide the ablation procedure. To reach this level of realism, real unipolar signals acquired at CHU Bordeaux are mapped on the simulated transmembrane potentials (see Fig. 2(a)). Each consecutive pair of electrodes thus computes one bipolar signal obtained by the difference of the two unipolar signals. The resulting bipolar signals acquired *in silico* are presented in Fig. 2(b). In the simulation, four electrodes are defined along the catheter tip and three bipolar signals driven by our MS model ("Cath1", "Cath2" and "Cath3") can be displayed.

Electro-Anatomical Mapping. Once inside the targeted cardiac chamber, the electro-anatomical mapping starts in order to localize the arrhythmic substrate, here an ectopic focus. As with a real mapping system, the position of the catheter is tracked and the endocardium is partially reconstructed when the catheter touches the heart wall. In other words, as soon as a collision between the catheter and the endocardium occurs, the intersected triangles of the navigation mesh are displayed. Noise is added, so that the reconstructed surface looks realistic. The surface resulting from the virtual reconstruction is shown in Fig. 3(b).

After reconstructing the endocardium, the simulation allows to build a map of activation times based on the extra-cellular measurements. When an ectopic stimulation starts while mapping, our framework computes local activation times, which are mapped on the endocardial surface. The activation times correspond to the elapsed time between the depolarization of the ectopic focus and the depolarization of the point currently in contact with the catheter. By measuring the extrasystolic activation times on the endocardium, the cardiology trainee can thus determine the exact location of the arrhythmia. The Fig. 3(c) captures this electro-anatomical mapping simulated during our virtual procedure.

RF Ablation. Once a target region has been identified, the cardiologist performs the RF ablation by heating the tissue using a RF (usually from 300 to 700 kHz) alternating current. This energy is delivered through an electrode in contact with the target tissue. When the temperature exceeds $60^{\circ}C$, denaturing of proteins

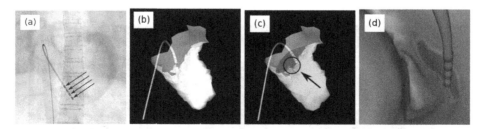

Fig. 3. (a) Fluoroscopic view of four electrodes (arrows) at the distal extremity of the catheter; (b) Reconstruction of the 3D RV endocardium; (c) Extrasystolic activation times with the ectopic focus (arrow); (d) Internal view of the RV where the red area underwent ablation

leads to a cellular death with coagulation necrosis. To be efficient, the temperature must nevertheless not exceed 100°C. After ablation, cardiac cells lose their electrical conductivity. As a consequence, a successful procedure assumes that the regions originally responsible for the electrical disorder are well electrically isolated.

In our simulation, the ablation step is modeled by a progressive decrease of the electrical conductivity inside the tetrahedra. A zero conductivity is associated to dead cardiac cells. The effect of ablation gradually propagates from edge to edge through the myocardium. The longer the ablation, the larger the ablation area. Our training simulator also allows to set the desired power of ablation, thus making the ablation process faster or slower. During the simulation, the ablation is triggered by one of both pedals provided with the tracking device. The tissue electrical conductivity is consequently updated in real-time. An illustration of an ablation scar is given in Fig. 3(d).

Stimulation. A stimulation can be used by the cardiologist to assess the success of the ablation procedure. When the catheter touches the heart wall, an electrical current can be delivered in the region of contact. Moreover, the current value of this catheter stimulus $J_{cath-stim}(t)$ is interactively set by the user. As for the ablation, the stimulation is triggered by the user using the pedals of the Mentice device. If no abnormal activity is detected during the stimulation process, the arrhythmic substrate is successfully ablated.

3   Results

Exploiting the power of both multithreading and GPU computing, we achieve a fully interactive simulation. Our simulation runs on a computer including an Intel Core i7 CPU and an NVidia GTX 580 GPU. The performance results are given in Table 1 regarding the different steps during the procedure. In this table, real-time ratio stands for the ratio of elapsed time over computation time. When the simulation is faster than real-time (ratio > 1), the computation can be slowed

Table 1. Performance results of the whole training framework

Position	Number of contacts		Training Simulator	
	Mean	[min-max]	Frames per second	Real-time ratio
Start	0	[0 - 0]	44.0	0.88
Vena cava	2	[2 - 2]	43.8	0.87
Entrance of atrium	5	[2 - 14]	31.6	0.63
Final	50	[41 - 65]	23.6	0.47

down to retain real-time. From Table 1, we first realize that the computational efficiency is strongly related to the number of contacts, i.e. to the navigation part. At the beginning of the procedure, only few contacts are detected whereas, once inside the heart, the catheter leans against the endocardium, thus decreasing the performance. This reveals that the navigation part tends to limit the overall performance when many contacts (over 50) are detected. 50 contacts implies 150 constraints which is substantial regarding the number of beams in our model (40 beams).

4   Conclusion

Table 1 shows that our training system runs between 1.1 and 2.3 times slower than real-time. Our hybrid architecture proves to be efficient so that the entire simulation remains interactive. Another crucial feature regarding our scope is that the interactions with the electrophysiology model do not affect the performances. As a consequence, this training simulator for cardiac RF ablation procedure already offers close to real-time performances and a high level of interactivity. Finally, a clinical evaluation is currently being conducted to assess the realism of our virtual training. This degree of validation would enlarge the field of application of our simulator.

References

1. Aliev, R., Panfilov, A.: A simple two-variable model of cardiac excitation. Chaos, Solitons and Fractals 7(3), 293–301 (1996)
2. Bertails, F., Audoly, B., Cani, M.-P., Querleux, B., Leroy, F., Lévêque, J.-L.: Super-helices for predicting the dynamics of natural hair. ACM Transactions on Graphics 25, 1180–1187 (2006)
3. Chiang, P., Zheng, J., Yu, Y., Mak, K., Chui, C., Cai, Y.: A vr simulator for intracardiac intervention. Computer Graphics and Applications 33(1), 44–57 (2013)
4. Dawson, S., Cotin, S., Meglan, D., Shaffer, D., Ferrell, M.: Designing a computer-based simulator for interventional cardiology training. Catheterization and Cardiovascular Interventions 51(4), 522–527 (2000)
5. Dequidt, J., Marchal, M., Duriez, C., Kerien, E., Cotin, S.: Interactive simulation of embolization coils: Modeling and experimental validation. In: Metaxas, D., Axel, L., Fichtinger, G., Székely, G. (eds.) MICCAI 2008, Part I. LNCS, vol. 5241, pp. 695–702. Springer, Heidelberg (2008)
6. Duriez, C., Cotin, S., Lenoir, J., Neumann, P.: New approaches to catheter navigation for interventional radiology simulation 1. Computer Aided Surgery 11(6), 300–308 (2006)
7. Johnson, D., Willemsen, P.: Accelerated haptic rendering of polygonal models through local descent. In: Haptic Interfaces for Virtual Environment and Tele-operator Systems, pp. 18–23 (2004)
8. Mansi, T., Pennec, X., Sermesant, M., Delingette, H., Ayache, N.: iLogDemons: A demons-based registration algorithm for tracking incompressible elastic biological tissues. International Journal of Computer Vision 92(1), 92–111 (2011)

9. Maron, B.J., Towbin, J.A., Thiene, G., Antzelevitch, C., Corrado, D., Arnett, D., Moss, A.J., Seidman, C.E., Young, J.B.: Contemporary definitions and classification of the cardiomyopathies. Circulation 113(14), 1807–1816 (2006)

10. Mitchell, C., Schaeffer, D.: A two-current model for the dynamics of cardiac membrane. Bulletin of Mathematical Biology 65, 767–793 (2003)

11. Rapaka, S., Mansi, T., Georgescu, B., Pop, M., Wright, G.A., Kamen, A., Comaniciu, D.: LBM-EP: Lattice-boltzmann method for fast cardiac electrophysiology simulation from 3D images. In: Ayache, N., Delingette, H., Golland, P., Mori, K. (eds.) MICCAI 2012, Part II. LNCS, vol. 7511, pp. 33–40. Springer, Heidelberg (2012)

12. Talbot, H., Marchesseau, S., Duriez, C., Sermesant, M., Cotin, S., Delingette, H.: Towards an interactive electromechanical model of the heart. Journal of the Royal Society Interface Focus 3(2) (April 2013)

13. Wang, F., Duratti, L., Samur, E., Spaelter, U., Bleuler, H.: A computer-based real-time simulation of interventional radiology. In: Engineering in Medicine and Biology Society, pp. 1742–1745. IEEE (2007)

A Virtual Reality System to Train Image Guided Placement of Kirschner-Wires for Distal Radius Fractures

Tian En Timothy Seah[1], Alastair Barrow[1], Aroon Baskaradas[2], Chinmay Gupte[2], and Fernando Bello[1]

[1] Simulation and Modelling in Medicine and Surgery,
Department of Surgery and Cancer
[2] Imperial College Healthcare NHS Trust
St. Mary's Hospital, Imperial College London, UK
a.barrow@imperial.ac.uk

Abstract. We present the design, development and initial user testing of a virtual reality simulator to train orthopaedic surgeons in the optimal placement of K-wires for fixation of distal radius fractures. Our platform includes 5 DOF haptic feedback to recreate the manual skill aspects of the drilling process, a 3D view of the anatomy and a controllable x-ray image. Once complete, the user is given an overview of their performance compared with the 'ideal placement' defined by an expert orthopaedic surgeon. The design goals based on analysis of the core steps in the procedure are presented, along with the technical implementation in terms of both haptic and graphical feedback. Preliminary user testing results are discussed, together with current limitations and planned future development.

1   Introduction

Current political, economic and ethical considerations have resulted in orthopaedic surgical trainees no longer being able to acquire the necessary skills from operating on patients alone. Virtual Reality (VR) surgical simulators offer a viable alternative and are poised to play an increasingly important role in the training of surgical skills.

We are currently investigating the potential of using VR simulation with haptic feedback to train orthopaedic surgeons in the optimal placement of Kirschner-wires (K-wires) for fixation of distal radius (wrist) fractures.

This is one of the first procedures that a trainee orthopaedic surgeon must master and is frequently performed in young and old patient populations after a fall onto an outstretched hand. It requires the placement of long, thin wires through the forearm, close to the wrist in order to fixate and bind the bone fragments in the correct position, stabilising the fracture while healing takes place. Accurate positioning of these wires based on an understanding of the particular fracture pathology and anatomical landmarks is essential and is performed using x-ray image guidance.

F. Bello and S. Cotin (Eds.): ISBMS 2014, LNCS 8789, pp. 20–29, 2014.
© Springer International Publishing Switzerland 2014

Research into bone drilling simulators with haptic feedback is well represented in the literature [1-5], but past work primarily focuses on algorithms, which attempt to accurately model the material removal process and tool-tissue contact. For the particular drilling skill this work considers (deep linear drilling), it is believed there is limited pedagogical value in recreating a physically accurate representation of the drilling mechanics, particularly as standard haptic devices cannot display the high stiffness or range of forces arising in orthopaedic drilling [6].

Instead, this work focuses on identifying the particular learning outcomes of the task and applying simple, but robust algorithms to give appropriate feedback to the user to promote learning. This 'user centric' methodology requires close interaction with expert surgeons when selecting and refining simulation parameters, though specific parameters in the simulation may not have direct, real-world, relevance.

We present the design stages and implementation to date, as well as initial results and user feedback for our distal radius fracture training simulator. The project is ongoing and the following will concentrate on the simulation and evaluation stages identified by the expert surgeons as most relevant to the procedure.

2   Task Analysis

Distal radius fractures are common [7]. Treatment options can be conservative or surgical, with the latter being divided mainly into plate fixation or K-wiring. K-wiring is of particular relevance to simulated training as it is a more basic skill that trainees need to master early and it is readily transferable to other orthopaedic procedures.

Surgery usually occurs under general anaesthetic with x-ray guidance. The patient is placed supine with the arm exposed for the surgeon to perform the procedure. The high-level steps of the procedure identified by an expert surgeon include:

- Identifying the relevant anatomy (radial styloid, fracture configuration)
- Deciding the entry point and trajectory for the first wire using x-ray
- Making a small skin incision at the entry point with the scalpel
- Advancement of the wire using a wire driver in the desired direction with continual x-ray screening
- Insertion of at least one more wire in another plane for maximal stability, once a satisfactory position of the first wire is achieved

The following were identified as desirable features and functionality through the initial task analysis and feedback from pilot testing with experts:

- Teaching and evaluating knowledge of anatomy and entry points (Fig. 1)
- Training the skill of using the x-ray and driving the wire simultaneously so that the surgeon can make dynamic minor alterations in wire position and angulation
- Realistic recreation of the variation in speed and resistance felt as the wire penetrates different parts of the bone, particularly the characteristic 'give' when the wire exits the bone alerting the surgeon to stop advancing the wire

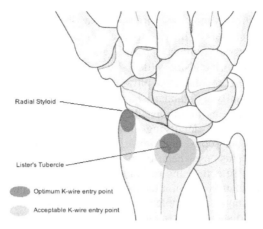

Radial Styloid

Lister's Tubercle

Optimum K-wire entry point

Acceptable K-wire entry point

Fig. 1. Optimal and acceptable entry sites as defined by an expert orthopaedic surgeon

The final point was of particular interest for a simulator with haptic feedback. During drilling, as the wire penetrates different parts of the bone, the speed and resistance felt by the operating surgeon changes: the proximal and distal cortices provide high resistance, whilst minimal resistance is experience between them. These subtle changes guide the surgeon to apply the appropriate amount of pressure, with a characteristic 'give' when the wire exits the bone alerting the surgeon to stop advancing the wire. Too little pressure and the wire will not advance or advance too slowly. Too much pressure and the wire may penetrate too far or cause the wire to bend or break.

3   Simulator Design

Unlike other orthopaedic procedures where cutting and shaping of the bone are important activities, the key manual skills involved in K-wiring are: positioning, angulation, depth and drill speed. This means that a simple, but flexible approach to recreating physical and haptic dynamics can be adopted rather than the more computationally expensive FEM or other volumetric techniques used in previous orthopaedic simulators [2, 3]. The angulation requirements defined in the task analysis and previous work suggesting both linear force and torque are equally strongly experienced during orthopaedic drilling [6], led to the integration of a 5 DOF haptic interface.

The simulator uses OpenSceneGraph[1] to render the 3D scene graphically and the Bullet Physics Library[2] for collision detection. A W5D[3] with a custom handle modelled as a common orthopaedic drill is used for haptic feedback (Fig. 2, Fig. 3). The haptic device is controlled using an impedance control paradigm, the kinematics of the W5D allow linear and rotation feedback to be controlled independently and are decoupled in the rendering as is described below. The non-actuated rotational degree of freedom is aligned with the long axis of the guidewire, as this has the least perceptible torque feedback and considered to have least relevance to learning outcomes.

---

[1] http://www.openscenegraph.org
[2] http://bulletphysics.org/
[3] http://www.entactrobotics.com/

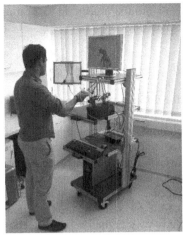 

Fig. 2. Simulation platform in use   **Fig. 3.** Close up view of haptic interaction

3.1   Surface Haptic Rendering

As the anatomy is represented using surface meshes, contact between the end point of the haptic tool and the outer surfaces of the anatomical models is performed using a standard triangular mesh haptic rendering technique [8]. The haptics thread runs at a 1000Hz update rate.

In addition to the point based surface interaction, the representation of the handpiece in the virtual world can slide over the guidewire to recreate the 'staggered' drilling as observed in [6]. This allows the user to adjust the length of guidewire exposed during drilling to gain better accuracy of the tip direction.

The additional torque DOF on the W5D is essential here to give a realistic impression of making contact with a distant point. The Haptic Interaction Point (HIP) follows the position of the physical handpiece. When colliding with a virtual object, a virtual point on the object surface (God Object, GO) is created. The force applied to the W5D can then be calculated based on x, the distance between the Haptic Interaction Point (HIP) and the GO:

$$\bar{F}_{GO} = k(GO - HIP)$$

k is the spring stiffness constant. The torque τ applied about the centre of mass is then calculated based on the length of the guidewire (Fig. 4):

$$\bar{\tau} = \bar{r} \times \bar{F}_{GO} \quad \text{where } \bar{r} = HIP\text{-}CoM$$

3.2   Percutaneous Haptic Rendering

Once the HIP has penetrated the bone (by the user depressing a button on the handpiece and pushing into the surface), the drilling algorithm takes over. The graphical drill render is locked to the GO, and as they progress into the object, an extra torque component is added to represent the surrounding material constraints (Fig. 5). The total torque then becomes:

$$\overline{\tau} = \overline{r} \times \overline{F}_{GO} + t\overline{\theta}$$

Where t is a torsional stiffness constant and $\overline{\theta}$ is the difference in orientation between the GO and the HIP.

On initial penetration of the surface, the angulation of the handpiece is unconstrained and drilling direction can be easily adjusted since there is minimal material surrounding the guide wire. As the guidewire proceeds into the bone, the torsional stiffness increases. In addition, the allowed range of angular movement gradually decreases. The orientation of the drill shown to the user on the screen is weighted between the original trajectory and the trajectory of the physical device. The deeper the HIP is inside the surface, the more it is weighted towards the original trajectory, limiting the range of angular movement. This occurs up to a predefined depth, after which the orientation and hence direction of drilling is fixed (Fig. 6).

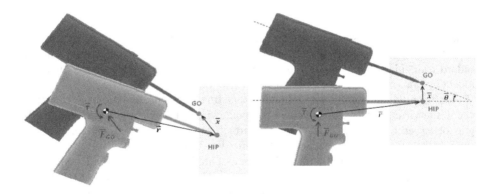

Fig. 4. Force and Torque acting on Centre of Mass (CoM) of device

Fig. 5. Additional torque component for percutaneous haptic rendering

The weighted trajectory of the GO, *i.e. what the user sees,* can be calculated by:

$$\overline{U} = \frac{d\overline{p} + (D - d)\overline{q}}{D}$$

Where d is current depth, D is the maximum depth at which angular movement is allowed, p is the stored trajectory, and q is the HIP orientation. U is updated in real time, but when drilling starts, it is locked such that the GO progresses in a straight line. When drilling is terminated, the most recent value of U is stored in p, to be used as a new reference trajectory for weighting subsequent re-angulations.

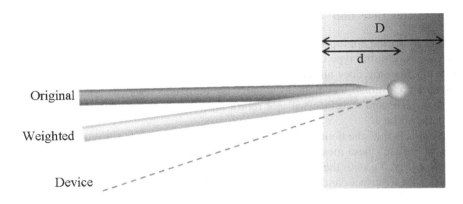

Fig. 6. Weighted Trajectory as a function of depth

Advancement of the guidewire is controlled by the user depressing a button on the handpiece and pushing forward. The force displayed to the user and the rate of ingress of the guidewire are both scaled by x, the distance between the GO and the HIP:

$$\bar{F}_{GO} = k\bar{x}$$
$$\bar{V}_{GO} = \sigma(\bar{F}_{GO} \cdot \bar{x})\bar{x}$$

Where V_{GO} is the rate of progress and has been determined by projecting the contact force onto the GO orientation. σ is the rate constant that represents the frequency of the update loop.

Additionally, the handpiece is made to vibrate with a differing pattern/amplitude under different conditions to increase realism. Each of the above parameters changes depending on the tissue currently being traversed: Cortical Bone, Trabecular Bone, Soft Tissue and Air (Table 1).

Table 1. Scaling of parameters according to tissue type

	k	α (vibration amplitude)	β (vibration Frequency)
Cortical	3.0 N/mm	High	50Hz
Trabecular	1.5 N/mm	Moderate	100Hz
Soft Tissue	1.0 N/mm	Low	200Hz
Air	0 N/mm	Lowest	250Hz

Of particular importance is the 'give' sensation when the guidewire transitions from a higher stiffness material to a lower (or air). This 'jerk' is recreated by discretely changing parameter values from layer to layer as indicated in Table 1.

3.3   Cortex Detection

In order to set the appropriate haptic and drilling feedback constants and to apply the 'give' sensation, the type of tissue the guidewire is passing through must be known. Since the anatomy does not consist of a volumetric representation, it is necessary to provide an additional technique to identify the particular tissue as collision tests between the HIP and a surface will not return a positive contact when the HIP is fully inside the surface.

To address this, the boundary of each tissue layer is defined by a polygon mesh. In order to 'inflate' these meshes and give them thickness, a sphere is used for collision detection (Fig. 7), with the radius of the sphere indicating how long it will remain in contact with a surface, and thus defining the cortical thickness, which is assumed to be between 2 mm to 5 mm [9].

When no contact is returned, a ray cast is performed. If the ray hits an even number of triangles, then it is determined to be outside the object. Otherwise, it is still inside the object.

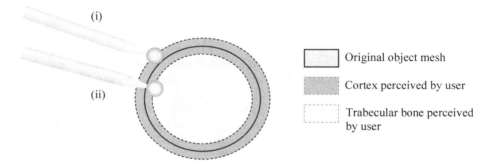

Fig. 7. Depiction of the sphere-surface collision used to 'inflate' the cortical surfaces

3.4   User Feedback

Once the user is satisfied with the guidewire placement, s/he can leave a K-Wire in place. At the end of the simulation, the user is given feedback on the placement accuracy of each K-wire. In addition, the radius is exposed and coloured spheres indicate the ideal entry and exit points. These entry/exit positions are manually defined and have been selected in consultation with expert surgeons and medical sources according to the type of surgery being simulated. Alignment of the K-Wires with these spheres will result in a higher score. Currently, ray casting is used to determine if the penetration points of the wires lie within the loci of the spheres. Other performance metrics recorded and presented to the user include: time taken, distance drilled, number of changes in direction and X-ray images taken (Fig. 8).

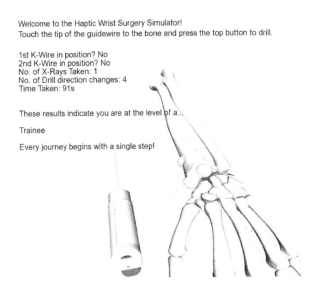

Welcome to the Haptic Wrist Surgery Simulator!
Touch the tip of the guidewire to the bone and press the top button to drill.

1st K-Wire in position? No
2nd K-Wire in position? No
No. of X-Rays Taken: 1
No. of Drill direction changes: 4
Time Taken: 91s

These results indicate you are at the level of a...

Trainee

Every journey begins with a single step!

Fig. 8. Example feedback presented to the user, basic metrics are displayed along with actual guidewire positions and 'idea' target positions

4   Pilot Study Procedure and Results

Following approval by the Local Research Ethics Committee, eleven orthopaedic surgeons (Junior – 6; Relatively Senior – 5) were invited to take part in the pilot study and early evaluation of the system. The study is ongoing and new participants are being recruited. Some of the initial qualitative and quantitative results are now presented

One particular early quantitative result of interest concerns the frequency of x-ray images taken per procedure. It is hypothesised that, once the platform reaches a sufficient level of sophistication, a difference will be found in x-ray frequency between junior and senior surgeons favouring a lower usage by those with more experience. The average frequency of x-ray images taken in actual surgery was estimated at 30-40 (currently anecdotal based on expert opinion). Descriptive statistics for both groups in the pilot study compared favourably with this estimate: Senior (M = 48.4, SD = 19.8, 95% CI [23.8, 73.0]) and Junior (M = 33.5, SD = 18.6, 95% CI [14.0, 53.0]). However, not only do the values indicate a higher rate in the senior groups, but a Mann-Whitney test (2-tailed, $U = 7.5$, $p = .2$) did not support any statistical difference between the two groups for the current data set. These tests will be repeated as the simulator is developed further.

Preliminary qualitative results indicate that reaction to the simulator is largely positive (21 comments from the 11 participants with 14 positive), including: *"Interesting way of learning a simple procedure.", "…like the haptic feedback!"*. Comments suggesting uncertainty or negative reactions included: *"…not sure how realistic it is.", "I'm sure bone would give more resistance."*

Many comments related to uncertainty in the use of the simulator or how it relates to performing the task in reality: *"Are we using two or three wires?"*, *"...happy with sphere location though I want to know if I can use the Kapandji technique..."* indicating subsequent revisions should focus on clearer instructional information embedding the user's experience in standard clinical procedure allowing for appropriate judgements and decision making to occur.

Initial responses to face validity were encouraging, particularly the physical interface: *"...handle felt right.", "This is just like the real one!"* with one subject indicating *"Handle is too light."* Which was an intentional design decision due to the limited force output of the haptic device. It was, however, observed the limited workspace of the haptic device caused some users difficulty when positioning the guidewire if fully extended from the hand piece.

The relevance of the simulator for more senior clinicians was questioned and in particular there were concerns about 'unlearning' a procedure they already knew how to perform on a real patient in order to score highly on the simulator: *"It is clear that more senior trainees find it frustrating as they already know how to perform the procedure and they're now having to relearn the procedure."*

As anticipated, there were a number of comments identified relating to aspects of the simulator likely to lie outside what can be simulated with the current setup: *"...no reduction possible.", "My left hand is useless...it doesn't feel natural...in real life I would manipulate the bone with my hand."*. These will need to be examined further and their pedagogical relevance considered with respect to the core intended outcomes of the simulator.

5   Discussion and Conclusions

A new simulation platform has been created to train surgeons in the placement of k-wires under x-ray guidance. The user interaction takes place through a 3D perspective view, an adjustable x-ray image and 5-DOF haptic feedback. The design decision was made to keep the haptic rendering simple and focus on the user experience rather than an underlying physically realistic model.

Current results are encouraging and the initial task analysis and pilot study have led to iterative improvements in the fidelity of the simulator. In particular, high importance was given to improving the perceptual difference between haptic feedback related to the different bone densities through which the guidewire passes. This has been developed and added to the simulation and is believed to address this training need, however, future evaluation is necessary to confirm this.

Whilst the evaluation phase of the project is ongoing, early results appear to suggest that the number of x-ray images taken per procedure, which was anticipated to be a discriminant between novice and expert surgeons is not only inappropriate, but in fact the relationship appears to have the opposite direction (junior surgeons using fewer x-rays).

A number of possible reasons for this results have been posited, including: senior surgeons who are "set in the ways" find adapting to the differences in the simulator challenging; senior surgeons may be more perfectionist aiming for a "perfect" result; the procedure in question may be too simple and only very junior surgeons will show a statistical difference for this measure; and, lastly as junior surgeons have performed fewer operations and experts no longer perform many trauma operations, there exists a "middle-cohort" who are more actively performing the procedure. More fundamentally, it may be that x-ray usage is not a useful discriminating factor for surgical experience in this particular procedure.

The work we have presented represents the initial design development and pilot studies for the simulator. We will use the results gathered at this stage to inform the next design revisions and recruit additional participants to further establish the face, content and construct validity of the simulator.

References

[1] Agus, M., Giachetti, A., Gobbetti, E., Zanetti, G., Zorcolo, A.: A multiprocessor decoupled system for the simulation of temporal bone surgery. Computing and Visualization in Science 5, 35–43 (2002)

[2] Blevins, N.H., Girod, S.: Visuohaptic simulation of bone surgery for training and evaluation (2006)

[3] Petersik, A., Pflesser, B., Tiede, U., Höhne, K.H., Leuwer, R.: Haptic volume interaction with anatomic models at sub-voxel resolution. In: Proceedings of 10th Symposium on Haptic Interfaces for Virtual Environment and Teleoperator Systems, HAPTICS 2002, pp. 66–72 (2002)

[4] Tsai, M.-D., Hsieh, M.-S., Tsai, C.-H.: Bone drilling haptic interaction for orthopedic surgical simulator. Computers in Biology and Medicine 37, 1709–1718 (2007)

[5] Vankipuram, M., Kahol, K., McLaren, A., Panchanathan, S.: A virtual reality simulator for orthopedic basic skills: A design and validation study. Journal of Biomedical Informatics 43, 661–668 (2010)

[6] Barrow, A., Akhtar, K., Gupte, C., Bello, F.: Requirements analysis of a 5 degree of freedom haptic simulator for orthopedic trauma surgery. Studies in Health Technology and Informatics 184, 43–47 (2012)

[7] Chung, K.C., Spilson, S.V.: The frequency and epidemiology of hand and forearm fractures in the United States. The Journal of Hand Surgery 26, 908–915 (2001)

[8] Melder, N., Harwin, W.S.: Extending the friction cone algorithm for arbitrary polygon based haptic objects. In: Proceedings of 12th International Symposium on Haptic Interfaces for Virtual Environment and Teleoperator Systems, HAPTICS 2004, pp. 234–241 (2004)

[9] Thompson, D.: Age changes in bone mineralization, cortical thickness, and haversian canal area. Calcified Tissue International 31, 5–11 (1980)

Using and Validating Airborne Ultrasound as a Tactile Interface within Medical Training Simulators

Gary M.Y. Hung[1], Nigel W. John[1,*], Chris Hancock[1], and Takayuki Hoshi[2]

[1] Bangor University, Bangor, UK
{g.m.y.hung,n.w.john,c.hancock}@bangor.ac.uk
[2] Nagoya Institute of Technology, Japan
hoshi.takayuki@nitech.ac.jp

Abstract. We have developed a system called UltraSendo that creates a force field in space using an array of ultrasonic transducers cooperatively emitting ultrasonic waves to a focal point. UltraSendo is the first application of this technology in the context of medical training simulators. A face validation study was carried out at a Catheter Laboratory in a major regional hospital.

Keywords: Airborne Ultrasound, Acoustic Radiation Pressure, Tactile, Medical Simulation, Palpation, Augmented Reality.

1 Introduction

Medical trainees typically gain experience by performing procedures on real patients whilst supervised by an expert. Porta [1] recently evaluated the willingness of patients to have a trainee performing a surgical procedure and found that they are more reluctant to allow a trainee to perform the operation in comparison to a senior clinician. This makes gaining consistent, frequent access to a significant number of patients for training purposes difficult. Medical simulators are one solution and offer a platform on which the trainee can practice for long periods without the worry of making mistakes on a real patient. Virtually created patients are computer generated and use different display technologies (often stereoscopic) to provide a visual representation. Unlike mannequins, virtual patients typically have no physical presence. However, most commonly performed procedures, such as a palpation, involves the clinician interacting with the patient using their hands or fingertips to conduct diagnosis or treatment. Therefore, it is important to simulate and stimulate the physical sense of touch when presenting a virtual patient to make the training session of sufficient fidelity to be beneficial.

A physical phenomena known as Acoustic Radiation Pressure (ARP) [2] is generated by focusing multiple ultrasonic transducers to create a force field. The force can be concentrated at a particular point in three-dimensional space and the width of the focal point can be as small as 5 mm in the transmission medium of air. This paper demonstrates the feasibility of using ARP from focussed airborne ultrasound to

---

[*] Corresponding author.

F. Bello and S. Cotin (Eds.): ISBMS 2014, LNCS 8789, pp. 30–39, 2014.
© Springer International Publishing Switzerland 2014

simulate common tactile sensations found on the body of a patient. We expand on out initial results [3] that focussed only on simulating a pulse like sensation, and provide the results of an evaluation of the system that was conducted at a major regional hospital.

2   Background and Related Work

Palpation is a common and necessary skill that all clinicians should be able to do. They may use their fingertips to feel a pulse, which is caused by the motion of the artery wall as the heart beats to circulate blood around the body. Some common palpable arteries are the carotid (neck), radial (wrist) and femoral (thigh). Examples of implementing a pulse in medical simulators include using a hydraulic pump to inflate a rubber tube embedded in a tray of silicone [4], or using a commercial force feedback joystick with a modified end effector [5].

A heart murmur is a condition where the valves of the heart show abnormal pulsations. A thrill sensation is also palpable and can be felt with the palm of the hand. The sensation is similar to a vibration but has a larger region of effect in comparison to an arterial pulse felt with the fingertips. An auscultation simulator [6] used a mannequin with embedded speakers to replicate the audible sounds of the heart. However, it is mainly designed to be used with a stethoscope and does not present a palpable thrill.

Other technologies have been used to provide tactile feedback. Shape memory alloys are typically used to achieve linear actuation of pin arrays [7, 8] but are known to suffer from slow response times, not sufficient to simulate a pulse rhythm. A pneumatic balloon array [9, 10] is an alternative to pin arrays but cannot achieve the same height as pins. Air tubes can also be used to create pressure on the pad of the finger [11] but requires the finger to be fully compressed against the valves to build up significant pressure, whereas if solenoids are used [12, 13] the devices tend to be too big. Finally, electrocutaneous stimulation could be achieved with electrodes [14] but requires a more complex control mechanism to display a desired sensation and may cause pain to the user.

2.1   Airborne Ultrasound

ARP is the acoustic pressure from a sound wave. From linear momentum, it can be explained as air particles pushing against the surface of an object (Fig.1). There will be many particles hitting against the surface and so the total force is an accumulation of the particles. The equation for radiation pressure on a surface [15] is given as:

$$P = \alpha \frac{p^2}{\rho c^2} \tag{1}$$

where p is total sound pressure from all ultrasonic transducers $[Pa]$, α is the constant of reflection coefficient, ρ is the density of the transmission medium $[kg/m^3]$ and c is the speed of sound in the transmission medium $[m/s]$. Equation (1) suggests it is possible to control the radiation pressure (P) as it is exponentially proportional to the sound pressure from the total number of ultrasonic transducers (p).

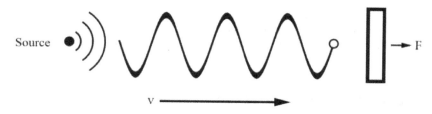

Fig. 1. The momentum of a particle transfers energy to the surface it hits. An accumulation of particles generates a sufficient amount of force that can be felt on the palm of a hand.

There are a few exemplar applications using airborne ultrasound for tactile feedback but this emerging technology has not yet been used in medical simulators. The first demonstration of using ARP was to co-locate a haptics sensation with the visual output from a holographic display giving the effect of making holograms touchable [16]. Later examples include a musical instrument with tactile feedback [17], allowing aerial input to a computer [18], adding tactile feedback to a handheld device for watching and feeling television content [19] and establishing a link between digital content and the physical world via an interactive table [20].

We believe that airborne ultrasound also offers the potential to simulate a tactile sensation such as a pulse or thrill sensation, and be able to simulate patients with different pulse rates or varying levels of strength in their pulse sensation due to age or body fat. Also a patient may exhibit arrhythmia, an abnormal condition featuring irregular pulse patterns, rate and rhythm. Airborne ultrasound offers greater controllability as modulating the amplitude of the control signals will directly influence the strength of the displayed sensation. A second benefit is airborne ultrasound's greater spatial resolution, which provides more control over where the sensation can be displayed i.e. the displayed sensation is not permanently fixed in one location. This will allow other scenarios to be simulated and provide more flexibility.

3   UltraSendo Configuration

A schematic of the UltraSendo hardware components is shown in Fig. 2. The PC contains the software for an Altera Field-Programmable-Gate-Array (FPGA) circuit board, the software runs indefinitely until switched off. The FPGA circuit board acts as the controller for UltraSendo and generates 40 kHz square wave signals driving the ultrasonic transducers. The pulsation or thrill effect is achieved by modulating the global duty cycle of this square wave. This technique is similar to pulse-width modulation by which we superimpose a periodic ON and OFF interval to the original 40 kHz wave. This technique allows us to simulate different pulse rates to represent different patients, different anatomical pulses or even irregular rhythms. The FPGA is connected to an intermediary distribution board that sends copies of the signal to each amplifier. Since there is only one signal pattern in the current implementation, it is not yet possible to re-position the focal point. If such capability is desired, however, the system design allows for a cost-effective upgrade by replacing the distribution board with another that feeds unique signals to the amplifiers.

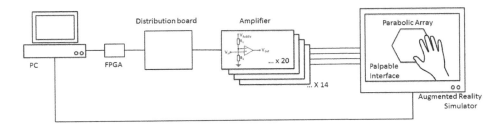

Fig. 2. Schematic of UltraSendo

Fourteen amplifier boards have been used, each consisting of twenty LM301AN operational amplifiers, providing one dedicated amplifying channel for each transducer. The boards are positioned in two stacks to save space. By making separate amplifier boards hosting small sets of twenty op-amps, it is possible to expand the system to a higher number of channels as required and thus increase the number of transducers when higher output is needed. The amplification circuit is a comparator that switches output from +15V to -15V. This polarity swing ensures the maximum displacement of the transducers diaphragm. Since the intermediate values in the digital signal are unnecessary for maximum output, they can be discarded. However, this also means it is not possible to modulate the intensity of the emitted beam by changing the amplitude of the original square wave signal.

A flat planar array of transducers would require phase shifts in the control signal to compensate for the different distances that the ultrasonic beams from each transducer have to travel to reach the focal point. UltraSendo uses a parabolic array so that the distance from each transducer to the focal point is the same (Fig 3). Thus the array can be driven with just one control signal that is copied and distributed to each transducer. Since all transducers are orthogonal to the focal point there is little power loss from beam divergence allowing a larger force to be achieved compared with using a flat array. The transducers in the parabolic array are arranged in the shape of a hexagon to fit as many transducers as possible into a given space. Combined with the use of smaller transducers (10 mm wide), UltraSendo is able to make use of more transducers than other implementations, e.g. Ciglar's parabloic array [17]. The resulting force output of UltraSendo was measured to be around 2.08 gf (gram-force) which surpasses the force output reported in [21] of 1.6 gf.

Existing airborne ultrasound devices are contact free and display the sensation as is i.e. feeling the raw acoustic waves. However, this sensation of touching air does not closely resemble that of touching the skin of a patient. The construction of UltraSendo has therefore taken a hybrid approach. For improved face validity, we use a simple membrane patch to represent the skin. This patch is structurally suspended horizontally above the parabolic array allowing the ultrasound to focus directly underneath the membrane. A flat membrane has been used to approximate the palpation and thrill sites. Although there will be some curvature on the surface of a real patient, the skin is flattened as the clinician applies force with their hand and fingers. Due to limitations in the maximum output force of airborne ultrasound, the selection of the membrane material needed consideration. The material cannot be too thick otherwise it will mask the pressure from the focal point. The material must also be non-porous

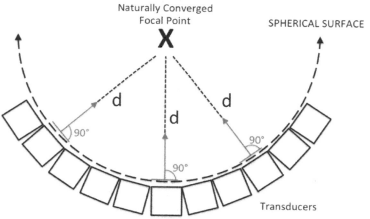

Fig. 3. Hexagonal parabolic array consisting of 271 ultrasonic transducers. All the ultrasonic beams will travel a distance of d to reach the focal point.

otherwise the focal point will not push against the surface and air particles by the ultrasound beam will bypass the surface through micro pores. The initial material that we are testing is polythene, a widely available thermoplastic polymer consisting of long hydrocarbon chains (commonly used for shopping bags). It is light and non-porous.

Finally UltraSendo was integrated into an augmented reality simulator (Fig.4). We used an approach pioneered in our PalpSim simulator [4], which uses chroma-key techniques to allow the trainees own hands to be superimposed onto the computer graphics rendering of the patient. UltraSendo's hardware components are therefore covered in blue cloth so that they can be masked out of the scene rendered on the computer monitor. A webcam is positioned underneath the monitor to provide a video feed of the trainees' hands. The trainee looks down at the monitor as if he/she is looking down at a patient.

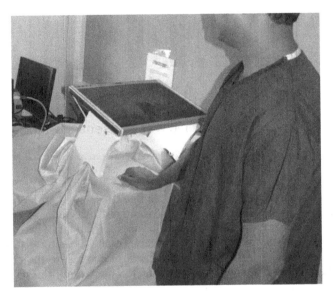

Fig. 4. One of the clinicians from Glan Clwyd Hospital evaluating UltraSendo. He can see his own hand placed on the palpable surface in the virtual patient displayed on the horizontally mounted monitor.

4   Face Validity Study

UltraSendo was taken to the Catheter Laboratory at the Ysbyty Glan Clwyd hospital in North Wales to collect expert user feedback. The laboratory is a major facility in the region for the treatment of patients with heart and chest conditions. Eleven clinicians made up of seven males and four females participated in this session. The majority of the participants specialised in Cardiology. They were asked to explore UltraSendo's palpable interface with their fingertips to search and feel for a pulse sensation. Then they were asked to explore the same palpable interface with their palm to feel for a thrill sensation (Fig.4). The strength of the force output was adjusted according to these different sensations, and to compensate for the distance to the skin of the anatomy that is causing the pulse or thrill (the heart is deeper inside the body, for example). After this interactive session, they completed a short questionnaire to rate how realistic they believed the sensation was as a pulse and thrill.

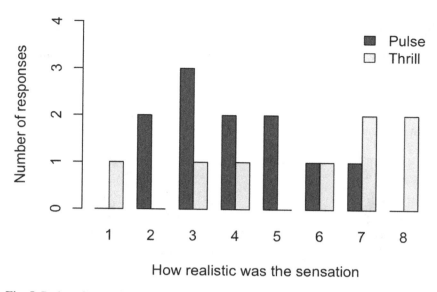

Fig. 5. Ratings for each interpretation of the displayed sensation as an arterial pulse and a thrill sensation

Referring to the bar plot (Fig.5) the distribution of responses for both the thrill and pulse sensation is shown. For the pulse sensation, UltraSendo received a mixed review with the feedback spanning the whole range of ratings. The majority of the responses are towards the middle left of the rating scale meaning that the sensation felt was not satisfactorily realistic enough. This is in contrast to the positive results that we obtained with the hydraulic system that we reported in [4]. For the thrill sensation, there is also a mixed review but responses do reside closer to the "very realistic" rating. Certainly the thrill sensation was received as more realistic than the pulse.

The clinicians were asked to indicate their years of experience in their specialty from the following options: 1-5, 5-10, 10-15 and +15 years and how often they palpate patients on a monthly basis. These parameters are plotted as a 3D graph (Fig.6). Whereas the sample size is too small to make any definite conclusions, some potential trends can be observed. A best fit regression plane is drawn for this plot to show the relationship between the parameters. The plane declines from left to right as years of experience and frequency of palpation increases. The junior clinicians are rating the fidelity of UltraSendo higher possibly because they have less experience of the sensation of a real palpation with which to compare against. The senior clinicians have been exposed to a wider range of forces and tactile sensations and thus could suggest the force generated by UltraSendo is apparently weaker than the real sensation.

Several comments from the participants were also noted. Some suggested that the force of the sensation needed to be increased to closer simulate the pulse sensation that the participant expects to find when palpating a real patient. A further observation made whilst watching the participants palpate UltraSendo was that almost all the participants initially depressed the membrane patch with substantial amount of force. The cardboard structure used for the UltraSendo prototype was not strong enough on occasions and could buckle. This is probably the reason why the participants could not

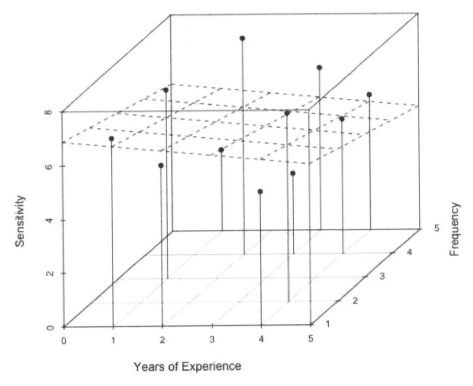

Fig. 6. 3D plot to show the relationship between the ability to detect the sensation against the participants' experience in their field

palpate with more strength. Another participant suggested the palpable surface could be improved by presenting a surface that felt like human skin to increase the realism of the simulator. The palpable surface is currently covered by the blue drapes to integrate the device into the augmented reality simulator and thus may not feel like touching the skin of a human.

5   Conclusions and Future Work

UltraSendo has enabled a tactile sensation to be achieved in a medical simulator for the first time using ARP. Our goal has been to provide a programmable tactile interface using an actuator that has no moving parts. The face validity is not yet sufficient to be able to deploy a production version of this technology and other tactile solutions will currently perform better. However, constructive feedback from participants in the face validation study highlighted the areas that need to be improved. In particular, the force from UltraSendo is weaker compared to palpating a real patient. This could be due to the small output force of the transducer array or the sturdiness of the physical infrastructure. For example, the palpable surface is particularly fragile as the volume underneath this surface is hollow. Thus the amount of force a user can apply to this surface is limited. Replacing the transmission medium of air with a denser medium such as water or gel can provide the physical support that the palpable surface requires in order to tolerate a higher amount of applied force. The propagation of ultrasound through mediums other than air have also been shown to generate a greater force magnitude. The textural sensation can also be improved by using a material that feels closer to human skin, such as elastomer, which can tolerate higher degrees of elastic deformation and thus is more resilient to damage compared to a thin sheet of polythene. This material can also be molded to represent the contours of a real patient – the focal point of the transmitted ultrasound can be adjusted for both curved and flat surfaces. The next phase of this research will therefore explore alternative ultrasound transmission mediums.

Currently UltraSendo displays the pulse/thrill sensation at a fixed focal point. This is because all 271 transducers are driven by the same control signal. To have the ability to relocate the focal point of airborne ultrasound, each transducer must have its own control signal. By replacing the Distribution board that the amplifier port is connected to, with another printed circuit that makes unique connections from the amplifier to the FPGA component, each transducer can be driven individually. This allows phase differences (delays in the signal) to be introduced and so the ability to change the location of the tactile effect anywhere within the region above the array of transducers. It would also be feasible to design a motion platform onto which the transducer array can be mounted and so change the location of the tactile effect by physically moving the array. This does, however, introduce more moving parts into the hardware and so introduce more reliability and robustness issues.

Changing the parameters such as the strength, location and the rhythm of the tactile sensation can simulate different patient profiles. For example a weak pulse suggests the patient may be elderly or has high body fat percentage. An irregular pulse rhythm could also indicate arrhythmia, a difficult condition to train the detection of as accessibility to real patients with this condition is limited. UltraSendo already supports the ability to present abnormal pulse rates and beat patterns. The next validation study will therefore also gather user feedback on how realistically UltraSendo can simulate different patient conditions.

Acknowledgements. We would like to thank our clinical collaborators: Professor Michael Rees, Cath Lab, Ysbyty Glan Clwyd; and Professor Derek Gould, Royal Liverpool Hospital. This research was supported by the Wales Research Institute of Visual Computing (RIVIC).

References

1. Porta, C.R., Sebesta, J.A., Brown, T.A., Steele, S.R., Martin, M.J.: Training surgeons and the informed consent process: routine disclosureof trainee participation and its effect on patient willingness and consentrates. Archives of Surgery 147(1), 57–62 (2012)
2. Iwatani, J.: Studies on Acoustic Radiation Pressure I. The Journal of the Acoustical Society of America 27(1), 198 (1955)

3. Hung, G.M.Y., John, N.W., Hancock, C., Gould, D.A., Hoshi, T.: UltraPulse - Simulating a Human Arterial Pulse with Focussed Airborne Ultrasound. In: 35th Annual International Conference of the IEEE EMBS, Osaka, Japan, vol. 550, pp. 2511–2514 (2013)
4. Coles, T.R., John, N.W., Gould, D.A., Caldwell, D.G.: Integrating haptics with augmented reality in a femoral palpation and needle insertion training simulation. IEEE Trans. on Haptics 4(3), 199–209 (2011)
5. Ullrich, S., Kuhlen, T.: Haptic palpation for medical simulation in virtual environments. IEEE Transactions on Visualization and Computer graphics 18(4), 617–625 (2012)
6. Takashina, T., Masuzawa, T., Fukui, Y.: A new cardiac auscultation simulator. Clin. Cardiol. 13(12), 869–872 (1990)
7. Hoshi, T., Abe, D., Shinoda, H.: Adding Tactile Reaction to Hologram. In: 18th IEEE International Symposium on Robot and Human Interactive Communication, pp. 7–11 (2009)
8. Kontarinis, D., Son, J., Peine, W., Howe, R.: A Tactile Shape Sensing and Display System for Teleoperated Manipulation. In: IEEE International Conference on Robotics and Automation, pp. 641–646 (1965)
9. Matsunaga, T.: 2-D and 3-D Tactile Pin Display Using SMA Micro-coil Actuator and Magnetic Latch. In: 13th International Conference on Solid-State Sensors, Actuators and Microsystems, pp. 325–328 (2005)
10. Santos-Carreras, L., Leuenberger, K., Retornaz, P., Gassert, R., Bleuler, H.: Design and Psychophysical Evaluation of a Tactile Pulse Display for Teleoperated Artery Palpation. In: International Conference on Intelligent Robots and Systems, pp. 5060–5066 (2010)
11. Wottawa, C., Fan, R.E., Lewis, C.E., Jordan, B., Culjat, M.O., Grundfest, W.S., Dutson, E.P.: Laparoscopic Grasper with an Integrated Tactile Feedback System. In: International Conference on Complex Medical Engineering, pp. 1–5 (2009)
12. Oakley, I., Kim, Y.: Combining Point Force Haptic and Pneumatic Tactile Displays. In: Proceedings of the EuroHaptics Conference, pp. 309–316 (2006)
13. Frisken-Gibson, S.F., Bach-y-Rita, P., Tompkins, W.J., Webster, J.G.: A 64-Solenoid, Four-Level Fingertip Search Display for the Blind. IEEE Trans. on Biomedical Engineering 34(12), 963–965 (1987)
14. Kajimoto, H., Kawakami, N., Tachi, S., Inami, M.: SmartTouch: Electric Skin to Touch the Untouchable. IEEE Computer Graphics and Applications 24(1), 36–43 (2004)
15. Hoshi, T., Takahashi, M., Iwamoto, T., Shinoda, H.: Noncontact Tactile Display Based on Radiation Pressure of Airborne Ultrasound. IEEE Trans. on Haptics 3(3), 155–165 (2010)
16. Sato, K., Tachi, S.: Design of Electrotactile Stimulation to Represent Distribution of Force Vectors. In: IEEE Haptics Symposium, pp. 121–128 (2010)
17. Ciglar, M.: An Ultrasound Based Instrument Generating Audible andTactile Sound. New Interfaces for Musical Expression, 19–22 (2010)
18. Hoshi, T.: Development of Aerial-Input and Aerial-Tactile-FeedbackSystem. In: World Haptics, pp. 569–573 (2011)
19. Alexander, J., Marshall, M.T., Subramanian, S.: Adding HapticFeedback to Mobile TV. In: ACM Conference of Human Factors, pp. 1975–1980 (2011)
20. Marshall, M., Carter, T., Alexander, J., Subramanian, S.: Ultra-Tangibles: Creating Movable Tangible Objects on Interactive Tables. In: Proceedings of the 30th International Conference on Human Factors in Computing Systems, pp. 2185–2188 (2012)
21. Hoshi, T., Iwamoto, T., Shinoda, H.: Non-contact Tactile Sensation Synthesized by Ultrasound Transducers. In: 3rd World Haptics, pp. 256–260 (2009)

Haptics Modelling for Digital Rectal Examinations

Alejandro Granados[1], Erik Mayer[2], Christine Norton[2], David Ellis[2],
Mohammad Mobasheri[2], Naomi Low-Beer[2], Jenny Higham[2],
Roger Kneebone[1], and Fernando Bello[1]

[1] Simulation and Modelling in Medicine and Surgery, Department of Surgery and Cancer
[2] Imperial College Healthcare NHS Trust
St. Mary's Hospital, Imperial College London, UK
a.granados@imperial.ac.uk

Abstract. Digital Rectal Examination (DRE) plays a crucial role for diagnosing anorectal and prostate abnormalities. Despite its importance, training and learning is limited due to their unsighted nature. Haptics and simulation offer a viable alternative for enhancing the learning experience by allowing the trainees to train in safety whilst trainers are able to assess competency. We present results of our geometrical, deformation and haptics modelling for two key anatomical structures obtained from patient specific MRI scans, namely the rectum and the prostate. Rectum *mobility* and *hardness* are modelled via a centreline consisting of control and structure points that are ruled by a mass-spring model based on elastic energy. Prostate *mobility*, *hardness*, *deformability* and *friction* are modelled via a surface model consisting of colliding spheres interconnected by springs with elongation, flexion and torsion properties. Clinical input and model fine-tuning was provided by three consultants from clinical disciplines that routinely perform DREs. Our approach is modular with scope to support additional palpable anatomical structures and the potential to be used as a teaching and learning tool for DRE.

Keywords: Digital Rectal Examination, Internal Examinations, Haptics Modelling, Prostate Cancer, Anorectal abnormalities, Deformation.

1   Introduction

Digital Rectal Examination (DRE) is recognised as a core skill to be taught as part of the medical curriculum. During DRE, the finger is inserted through the back passage to diagnose anorectal and prostate abnormalities. Among the structures that may be palpated are: the prostate (lobes and medium sulcus), seminal vesicles (only occasionally), anus (anal canal, inter-sphincter groove, anorectal junction), rectum (rectal ampulla), peritoneum (only occasionally), coccyx, and ischial spine (only possible in thin individuals). DRE plays a key role in the early diagnosis of anorectal [1,2] and prostate [3] abnormalities. However, teaching and learning of DRE, as well as competency assessment, are limited. Current benchtop models are unable to reproduce the wide range of normal and abnormal findings. Practicing on patients has its own limitations as they may be unwilling to be examined by an inexperienced trainee. The

F. Bello and S. Cotin (Eds.): ISBMS 2014, LNCS 8789, pp. 40–49, 2014.
© Springer International Publishing Switzerland 2014

unsighted nature of DRE renders both traditional approaches ineffective in terms of formative or summative assessment. Simulation offers obvious benefits by improving the learning experience through timely feedback and its ability to systematically expose trainees to a wide range of normal and abnormal anatomy.

Previous work on modelling organs for unsighted examinations includes a training tool for the diagnosis of prostate cancer [4], haptic feedback models for organ-to-organ interaction [5], bovine reproductive tract [6] and gynaecological examination [7]. The underlying models presented by these authors tend to be simplistic in terms of geometry, deformation and haptic interaction.

We propose a flexible and modular modelling approach based on patient-specific data, capable of supporting normal variability and common abnormalities of rectum and prostate, as well as the interaction of the examining finger with palpable objects through the constraining rectal walls. Models for the finger, rectum and prostate are presented. Firstly, the finger, placed in a thimble attached to a haptic device, is represented by a sphere that collides with the coccyx, rectum and prostate models. Rectum *mobility* and *hardness* are modelled via a centreline consisting of a set of control points supported by structure points. Control and structure points are interconnected with elastic links via a mass-spring model based on elastic energy. *Mobility* of the rectum is achieved by assigning stiffness values to these links. Haptics modelling of the rectum is based on the interpolation of neighbouring control points and the penetration depth inside the rectum walls. *Hardness* of the rectum is controlled by defining a stiffness constant in the reaction force. Prostate *mobility*, *hardness*, *deformability* and *friction* are modelled via a surface model consisting of colliding spheres that are interconnected by springs with elongation, flexion and torsion properties. *Mobility* of the prostate is controlled by setting a spring constant based on the resting position. *Deformability* of the prostate is defined by stiffness properties of the springs and the displacement of surface spheres towards the centre of gravity, with lumps modelled by stiff links and fixed spheres. Collision detection is determined based on penetration depth. *Hardness* of the prostate is obtained by computing reaction forces based on the surface normal, whereas *friction* of the prostate is modelled via friction planes / arcs. Geometrical modelling is described first, followed by deformation modelling and haptics modelling. Model fine-tuning and clinical validation are then discussed, with conclusions and future work presented at the end.

2   Geometrical Modelling

Ethics approval was obtained from the NHS National Patient Safety Agency Research Ethics Committee to recruit ten male volunteers (age range 18 to 65 years with informed consent and no known genito-urinary or colorectal disease) who were MRI scanned (GE Medical Systems Discovery MR 750 3.0T MRI scanner with resolution of 0.625x0.625x0.6mm scanning the pelvic region). ITK-SNAP (www.itksnap.org) was used to segment the anatomical structures of interest. Active contours semi-automatic segmentation was used for the bladder and coccyx. Manual segmentation was preferred for the rectum and prostate. Meshes were refined using MeshLab (meshlab.sourceforge.net) Two-Step Smooth and Laplacian Smoothing filters (Fig 1).

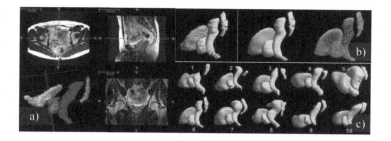

Fig. 1. Anatomical modelling: a) Segmentation of bladder, prostate, rectum and coccyx; b) Segmentation, smoothing and decimation; c) Ten patient-specific datasets

2.1   Rectum

Rectum walls are thin and deformable, allowing the palpation of surrounding structures such as the prostate, while constraining finger movement during examination. The geometrical model of the rectum consists of a discrete centreline defined by a set of control points $\mathbf{x}_{cp_i}$ with radius r_{cp_i} aligned to the rectum mesh. Each control point is surrounded by four fixed structure points $\mathbf{x}_{sp_i}$ (Fig 2.a).

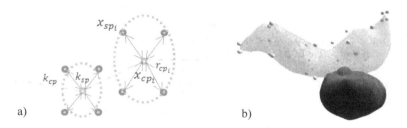

Fig. 2. Centreline construction based on a patient-specific 3D mesh of the rectum: a) centreline model consisting of a series of control points surrounded by structure points. b) Sample rectum model with a centreline comprising 6 control points and 24 structure points.

2.2   Prostate

Key considerations for modelling the prostate include the ability to assess prostate size, shape, amount of tissue deformation through applied pressure, and the presence of surface irregularities such as lumps. The geometrical model of the prostate consists of a decimated 3D surface mesh (Fig. 3.a) and a collection of spheres (Fig. 3.b). A non-fixed sphere i is created for each vertex of the mesh with equal radius r_i. Spheres are aligned to the surface of the mesh by displacing the centre $\mathbf{x}_i$ of each sphere along the opposite direction of the vertex normal $\mathbf{n}_i$ a distance equal to the radius r_i. Using this approach, a lump on the surface of the prostate may be generated by creating a fixed sphere with a radius larger than r_i. The prominence of the lump is controlled by the applied displacement along the opposite direction of the vertex normal $\mathbf{n}_i$.

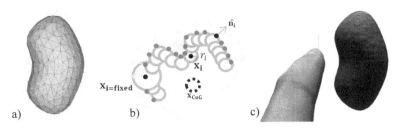

Fig. 3. a) Prostate 3D mesh (252 vertices, 500 faces); b) Non-fixed spheres surface model constructed from mesh vertices; a lump is represented with a larger fixed sphere; c) Sample spheres prostate model

3   Deformation Modelling

3.1   Rectum

A mass-spring model based on elastic energy U was implemented to constrain the movement of the control points forming the centreline. Each control point is linked to its adjacent control point(s) by springs with stiffness k_{cp} and to its four structure points by springs with stiffness k_{sp} (Fig. 2.a). *Mobility* of the rectum is achieved by assigning stiffness values to these links (Table 1). The force at any point of a spring is computed based on the negative derivative of the potential energy stored in the spring over displacement, i.e. from the spring resting length L_0 to its current length L (Eq. 1-2). We computed the internal force $\mathbf{F_i}$ of a control point i as the sum of all these forces from springs connected between control point i and neighbouring control/structure points j (Eq. 3).

$$U = \tfrac{1}{2}k(L-L_0)^2 \tag{1}$$

$$\mathbf{F} = -\nabla_x U \tag{2}$$

$$\mathbf{F_i} = \sum_j -k(L_{ij} - L_{0_{ij}})\tfrac{\mathbf{x_i}-\mathbf{x_j}}{L_{ij}} \tag{3}$$

External forces $\mathbf{F_{cp}}$ applied to control points are computed based on a deformation factor c_r and the collision response between the finger and the rectum walls represented as the negative of force $\mathbf{F_w}$, (Eq. 4, 8, 9).

$$\mathbf{F_{cp}} = -c_r * \mathbf{F_w} \tag{4}$$

3.2   Prostate

We used a skeleton model similar to the Chai3D GEL dynamics model [8], where a set of nodes with mass, radius and damping (linear and angular) are interconnected by springs with elongation k_E, flexion k_F and torsion k_T stiffness constants. *Deformability* of the prostate is defined by specifying elongation and flexion properties for these springs (Table 1). The external force $\mathbf{F_i}$ applied to a surface node i consists of a force $\mathbf{F_{m_i}}$ due to prostate movement and a force $\mathbf{F_{p_i}}$ due to palpation, multiplied by a deformation factor c_p (Eq. 5).

$$\mathbf{F_i} = c_p * (\mathbf{F_{m_i}} + \mathbf{F_{p_i}}) \tag{5}$$

$$\mathbf{F_{m_i}} = k_m * \left(\mathbf{x_{i_{res}}} - \mathbf{x_i}\right) \tag{6}$$

$$\mathbf{F_{p_i}} = -k_d * (r_{hd} + r_i - |\mathbf{x_{hd}} - \mathbf{x_i}|) * (\widehat{\mathbf{x_{hd}} - \mathbf{x_{CoG}}}) \tag{7}$$

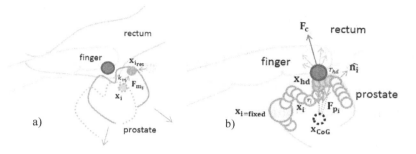

Fig. 4. Prostate modelling: a) external forces $\mathbf{F_m}$ for a moving prostate, and b) external forces $\mathbf{F_{p_i}}$ applied to palpated nodes with collision response $\mathbf{F_c}$

The prostate is allowed to move by computing a spring force $\mathbf{F_{m_i}}$ between the current position $\mathbf{x_i}$ of a surface node and its resting position $\mathbf{x_{i_{res}}}$ subject to stiffness k_m for all nodes comprising the prostate (Fig. 4.a and Eq. 6). *Mobility* of the prostate is controlled by specifying a value for this spring constant (Table 1). The force $\mathbf{F_{p_i}}$ due to palpation of a surface node i is computed as a spring force with stiffness k_d based on the penetration depth of colliding spheres in the direction towards the centre of gravity $\mathbf{x_{CoG}}$ of the prostate (Fig. 4.b and Eq. 7). *Hardness* of the prostate is controlled by specifying a value for this stiffness constant (Table 1).

4   Haptics Modelling

4.1   Rectum

Collision detection consists of three steps: first, the nearest neighbouring control point $\mathbf{x_{cp}}$ is determined based on the closest distance d_{min} to the finger, represented by the sphere with centre $\mathbf{x_{hd}}$ and radius r_{hd}. Then, the second nearest neighbouring control point $\mathbf{x_{ncp}}$ is found depending on whether $\mathbf{x_{hd}}$ lies below or above the plane formed by the closest control point $\mathbf{x_{cp}}$ and its normal $\mathbf{n_{cp}}$ (Fig. 5.a). Secondly, the centre-line position $\mathbf{x_i}$ and radius r_i are obtained through linear interpolation between the nearest neighbouring control points (Fig. 5.b). Lastly, the penetration depth of the finger represented by the sphere with centre $\mathbf{x_{hd}}$ and radius r_{hd} against the rectum wall with centre $\mathbf{x_i}$ and radius r_i is computed by adding the distance between both centres to r_{hd} and subtracting r_i (Eq. 8).

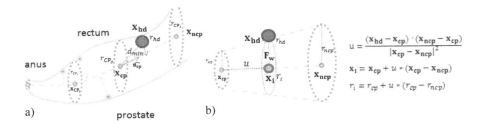

Fig. 5. Modelling of the rectum: a) Obtaining nearest neighbouring control points. b) Interpolation of centreline position $\mathbf{x_i}$ and radius r_i and computation of reaction force.

Collision response is expressed as the force $\mathbf{F_w}$ applied to the finger. $\mathbf{F_w}$ is obtained depending on whether the interpolated radius r_i is larger or smaller than the radius r_{hd} of the sphere representing the finger. If larger, $\mathbf{F_w}$ is computed as a spring force based on the penetration depth and stiffness constant k_w (Eq. 8), which defines the *hardness* of the rectum (Table 1). If smaller, $\mathbf{F_w}$ is computed as a spring force between the sphere representing the finger and the interpolated point $\mathbf{x_i}$ (Eq. 9). The section between the first ($\mathbf{x_{cp=0}}$) and third control points ($\mathbf{x_{cp=2}}$) corresponds to the narrow anal canal (Fig. 2.b left side of rectum model). This section is naturally narrower than the radius of the sphere representing the finger. If the sphere $\mathbf{x_{hd}}$ is within this section, then $\mathbf{F_w}$ in Eq. 8 or Eq. 9 is multiplied by a constant c_{ac}.

$$\mathbf{F_w} = k_w * (r_{hd} + |\mathbf{x_i} - \mathbf{x_{hd}}| - r_i) * (\widehat{\mathbf{x_i} - \mathbf{x_{hd}}}) \qquad \text{if } r_i > r_{hd} \qquad (8)$$

$$\mathbf{F_w} = k_w * |\mathbf{x_i} - \mathbf{x_{hd}}| * (\widehat{\mathbf{x_i} - \mathbf{x_{hd}}}) \qquad \text{if } r_i \leq r_{hd} \qquad (9)$$

4.2   Prostate

Collision detection consists of iterating through all the surface nodes and computing the penetration depth between the sphere representing the finger ($\mathbf{x_{hd}}$ and r_{hd}) and the spheres representing the surface of the prostate ($\mathbf{x_i}$ and r_i), by subtracting the distance between the centres of the spheres from the combined radii ($r_{hd} + r_i$).

Collision response is expressed as the force $\mathbf{F_c}$ applied to the finger. $\mathbf{F_c}$ is obtained by adding the forces resulting from contact with non-fixed spheres, contact with fixed spheres (lumps) and forces due to friction $\mathbf{F_{friction}}$ (Eq. 10). Forces while palpating non-fixed spheres are computed as spring forces with stiffness k_d, based on the penetration depth in the direction of the node normal $\mathbf{n_i}$ (Eq. 11). Forces while palpating fixed spheres are computed as spring forces with stiffness k_d based on the penetration depth in the direction of the point of contact on the surface of the fixed sphere, multiplied by a factor c_f (Eq. 12). This results in the palpation of a hard sphere (e.g. lump).

$$\mathbf{F_c} = \mathbf{F_{non-fixed}} + c_f * \mathbf{F_{fixed}} + \mathbf{F_{friction}} \qquad (10)$$

$$\mathbf{F_{non-fixed}} = \Sigma_i k_d * (r_{hd} + r_i - |\mathbf{x_{hd}} - \mathbf{x_i}|) * \widehat{\mathbf{n_i}} \qquad (11)$$

$$\mathbf{F_{fixed}} = \Sigma_i k_d * (r_{hd} + r_i - |\mathbf{x_{hd}} - \mathbf{x_i}|) * (\widehat{\mathbf{x_{hd}} - \mathbf{x_i}}) \qquad (12)$$

$$\mathbf{F_{friction}} = \Sigma_i k_f * (\mathbf{x_{GO_i}} - \mathbf{x_{SO}}) \qquad (13)$$

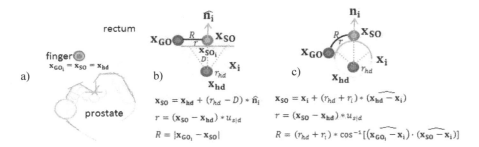

rectum

finger
$x_{GO_i} = x_{SO} = x_{hd}$

a)

prostate

b)

$\hat{n}_i$

x_{GO} —R— x_{SO}
r
x_{SO_i}
D:
x_i
r_{hd}
x_{hd}

$x_{SO} = x_{hd} + (r_{hd} - D) * \hat{n}_i$

$r = (x_{SO} - x_{hd}) * u_{s|d}$

$R = |x_{GO_i} - x_{SO}|$

c)

n_i

x_{GO}
R
r
x_{SO}
x_i
r_{hd}
x_{hd}

$x_{SO} = x_i + (r_{hd} + r_i) * (\widehat{x_{hd} - x_i})$

$r = (x_{SO} - x_{hd}) * u_{s|d}$

$R = (r_{hd} + r_i) * \cos^{-1}[(\widehat{x_{GO_i} - x_i}) \cdot (\widehat{x_{SO} - x_i})]$

Fig. 6. Friction modelling: a) Hybrid model consisting of friction planes and friction arcs (orange); b) Friction cone of a friction plane and c) Friction cone of a friction arc

Friction $\mathbf{F_{friction}}$ between colliding spheres $\mathbf{x_{hd}}$ and $\mathbf{x_i}$ is modelled as a spring force with stiffness k_f (Eq. 13). *Friction* of the prostate is controlled by specifying a value for this stiffness constant (Table 1). $\mathbf{F_{friction}}$ is based on the position of a God-Object $\mathbf{x_{GO}}$ and a Surface-Object $\mathbf{x_{SO}}$, together with the friction cone r formed by $\mathbf{x_{SO}}$ and the sphere $\mathbf{x_{hd}}$ representing the finger, as well as a static (u_s) or dynamic (u_d) friction constant [9]. We follow a hybrid approach that uses a series of friction planes in the case of non-fixed nodes, and a series of arcs in the case of fixed nodes (e.g. lumps) as illustrated in Fig 6.a. When the finger is palpating a non-fixed node, the position of the Surface-Object $\mathbf{x_{SO}}$ is the closest point on the plane formed by a surface point at the vertex node $\mathbf{x_{SO_i}}$ and its normal $\mathbf{n_i}$, a distance r_{hd} above the plane to ensure $\mathbf{x_{SO}}$ remains on the surface (Fig. 6.b). Alternatively, when the finger is palpating a rigid node, $\mathbf{x_{SO}}$ is set at a distance r_{hd} above the surface of the rigid sphere (Fig. 6.c). For both friction planes and friction arcs, R is then defined as the distance between $\mathbf{x_{SO}}$ and the previous $\mathbf{x_{GO}}$. For friction planes, R is the linear distance, whereas for friction arcs R is the great-circle distance. If this distance R is larger than the friction cone r, then static friction is replaced by dynamic friction as in [9].

5   Model Implementation and Validation

Ogre3d and HAPI were integrated to provide a common modelling framework. 3D meshes are parsed and loaded into Vertex and Index Buffer Objects in Ogre3d. Haptics is rendered either through haptic shapes and surfaces, or haptic effects in HAPI. An internal view of these meshes along with their modelling is shown in Fig. 7.

Three consultants from different clinical disciplines (Colorectal Surgery – CR, Urology – UR and Nursing – NP) were recruited to participate in the study. An initial validation of the models was conducted through a semi-structured interview and practical session interacting with the virtual models to determine an acceptable range of values for the various model parameters of healthy anatomy. Model parameters were grouped into behavioural categories for ease of interaction and interpretation. They were fine-tuned by participants whilst interacting with the models through the haptic device using slider bars with values ranging from -1 to 1. Categories and corresponding range of allowed parameter values for the rectum were mobility (k_{cp} [0,400], k_{sp} [0,200]) and hardness (k_w[0,120]), whilst for the prostate were mobility (k_m [0,60]), hardness (k_d [0,60]), deformability (k_E [0,200], k_F [0,0.2]) and friction (k_f [0,40]).

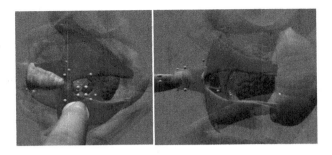

Fig. 7. DRE Simulator: Different views of virtual anatomy and examining finger with the interpolation of rectum centreline shown in pink and the prostate model shown as green spheres

The rectum was configured with six control points, total mass of 140g. Multiplier factors of 5.0, 10.0 and 5.0 were used for the stiffness k_{sp} of control points $\mathbf{x_{cp=0}}$, $\mathbf{x_{cp=1}}$ and $\mathbf{x_{cp=2}}$ in order to make this section (anal canal) more difficult to move. A deformation factor c_r equal to 6.0 was defined. Table 1 shows that UR and NP chose similar parameter values, while CR specified a less mobile and softer rectum. We also observed that, if mobility parameters decreased (more mobile), hardness parameters increased (harder) and vice versa. Two consultants (CR, NP) agreed that the rectum model was realistic (4 on a 5-point Likert scale) whilst the third consultant (UR) was neutral (3 on a 5-point Likert scale) (Fig. 8).

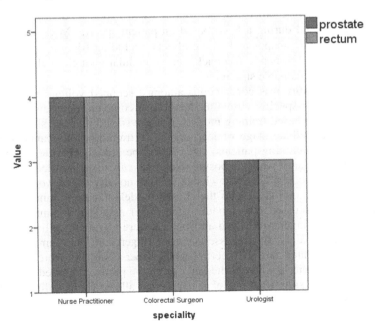

Fig. 8. Subjective feedback of prostate and rectum models on a 5-point Likert scale

The prostate was modelled using 5mm-radius surface spheres with a weight of 2g. k_T remained constant with a value of 0.1. Linear damping of 3.0 and angular damping of 1.0 were configured, whilst the deformation factor c_p and the fixed-node factor c_f were set to 1.0 and 3.0, respectively. Static friction coefficient u_s and dynamic friction coefficient u_d were set to 0.5 and 0.3, respectively. The deformation model was scaled up from the haptics model by a factor of 2.475 to guarantee stability in the integration. There was agreement amongst all three participants on mobility and between UR and NP on hardness, with no agreement for deformability (Table 1). Two consultants (CR, NP) agreed that the prostate model was realistic (4 on a 5-point Likert scale), whilst the third consultant (UR) was neutral (3 on a 5-point Likert scale) (Fig. 8).

Table 1. Experimental parameter values for rectum and prostate models (healthy anatomy)

Prostate		*CR*	*UR*	*NP*
Mobility	k_m	32	30	30
Hardness	k_d	16	36	33
Deformability	k_E	107	10	49
	k_F	0.11	0.01	0.05
Friction	k_f	16	14	29

Rectum		*CR*	*UR*	*NP*
Mobility	k_{cp}	201	85	90
	k_{sp}	101	42	45
Hardness	k_w	62	85	97

6   Discussion and Conclusions

We have presented our geometrical, deformation and haptic modelling approach to key structures palpated during a DRE based on patient-specific MRI scans. The approach combines haptic shapes and haptic effects, including friction, to generate the required type and level of haptic feedback, and is modular enough to allow the incorporation of additional palpable structures.

Anatomical variability was conspicuous amongst scanned volunteers, confirming the potential of patient-specific virtual models to overcome an important limitation of traditional mannequin-based training models. It is envisaged that our simulator could be used at an intermediate stage of learning, after training with mannequin-based models and before examining patients, since it will be able to provide feedback and assessment of performance whilst exposure to clinical cases is controlled.

In relation to the models, we acknowledge that the novelty of the work lies in their integration and application area, with the chosen models having a number of limitations that are now discussed. The model of the finger as a sphere simplifies the adjustment of different finger widths and allows for sphere-to-sphere collision detection. However, it does not allow for the assessment of fingerprint scale lumps/polyps and forces are only generated at the fingertip. The former could be overcome with the inclusion of tactile devices, whereas the latter might be partly addressed via the application of torques. The simplified model of the rectum allows for the adjustment of the centreline depending on patient position (knees towards chest) and for defining sections that have less mobility than others (sphincters). Touching the prostate and coccyx through rectum walls results in the summation of force effects from each organ. However, the geometrical model of rectum is only an approximation to the real anatomy. Lastly, the surface sphere model of the prostate is purposely simple to support efficient collision detection with the finger. The links interconnecting spheres allow for an easy configuration of region-based mechanical properties. A surface model was chosen as opposed to a volumetric model to avoid small elements. Lumps with cancer are easily represented by fixed spheres of different sizes. Finger fall-through problems are avoided by displacing spheres towards the centre of gravity and a hybrid friction model helps not only to avoid the finger sliding away, but also to detect lumps more easily. The prostate model does not scale well and we are unable to model cancer that is not located on the surface. These models have shown to be stable, suitable for real-time haptic interaction and easy to experimentally fine-tune by indicating the permissible amount of displacement, their softness or hardness, and the amount of deformation resulting from palpation. After fine-tuning, the behaviour of the models was judged to be from somewhat realistic to realistic (3 − 4 on a 5-point Likert scale).

Finger dimensions, patient-specific geometries and sets of patient-specific mechanical parameters can be combined and used to create different clinical scenarios. A learning scenario would randomly load a patient and select a set of parameter values for the rectum and prostate models that would represent either healthy organs or include pathologies (e.g. a number of lumps with specific size, cancer, enlargement).

Planned future work includes the modelling of the initial finger insertion phase through the anal canal, a model for the sphincters and a model for the anorectal junction. The use of haptic textures to incorporate anorectal abnormalities such as fissures and polyps will also be explored. Further validation of the rectum and prostate models based on soft tissue compliance, as well as additional experimental sessions with an increased number of experts from the relevant clinical disciplines.

References

1. Bharucha, A.E., Rao, S.S.C.: An Update on Anorectal Disorders for Gastroenterologists. Gastroenterology 146, 37–45 (2014)
2. Wong, R.K., et al.: The digital rectal examination: a multicenter survey of physician's and students' perceptions and practice patterns. Am. J. Gastroenterol. 107, 1157–1163 (2012)
3. American Cancer Society. Prostate Cancer: Early Detection (2013)
4. Burdea, G., et al.: VR-based Training for the Diagnosis of Prostate Cancer. IEEE Transactions on Biomedical Engineering 46(10), 1253–1260 (1999)
5. Kuroda, Y., et al.: Interaction model between elastic objects for haptic feedback considering collisions of soft tissue. Computer Methods and Programs in Biomedicine, 216–224 (2005)
6. Baillie, S., et al.: Validation of a Bovine Rectal Palpation Simulator for Training Veterinary Students. Studies in Health Technology and Informatics, 33–36 (2005)
7. dos Santos Machado, L., de Moraes, R.M.: VR-Based Simulation for the Learning of Gynaecological Examination. In: Pan, Z., Cheok, D.A.D., Haller, M., Lau, R., Saito, H., Liang, R. (eds.) ICAT 2006. LNCS, vol. 4282, pp. 97–104. Springer, Heidelberg (2006)
8. Conti, F., et al.: Chai3D: An Open-Source Library for the Rapid Development of Haptic Scenes. IEEE World Haptics (2005)
9. Barrow, A.: A dynamic Virtual Environment for Haptic Interaction. U. of Reading (2006)

Patient-Specific Meshless Model
for Whole-Body Image Registration

Mao Li[1], Karol Miller[1,2], Grand Joldes[1], Ron Kikinis[3], and Adam Wittek[1]

[1] Intelligent Systems for Medicine Laboratory,
School of Mechanical and Chemical Engineering
The University of Western Australia, Crawley-Perth, Australia
[2] Inst. of Mechanics and Advanced Materials,
Cardiff School of Engineering, Cardiff University, Cardiff, UK
[3] Surgical Planning Laboratory, Brigham and Women's Hospital,
Harvard Medical School, Boston, MA, USA

Abstract. Non-rigid registration algorithms that align source and target images play an important role in image-guided surgery and diagnosis. For problems involving large differences between images, such as registration of whole-body radiographic images, biomechanical models have been proposed in recent years. Biomechanical registration has been dominated by Finite Element Method (FEM). In practice, major drawback of FEM is a long time required to generate patient-specific finite element meshes and divide (segment) the image into non-overlapping constituents with different material properties. We eliminate time-consuming mesh generation through application of Meshless Total Lagrangian Explicit Dynamics (MTLED) algorithm that utilises a computational grid in a form of cloud of points. To eliminate the need for segmentation, we use fuzzy tissue classification algorithm to assign the material properties to meshless grid. Comparison of the organ contours in the registered (i.e. source image warped using deformations predicted by our patient-specific meshless model) and target images indicate that our meshless approach facilitates accurate registration of whole-body images with local misalignments of up to only two voxels.

1 Introduction

Registration of medical radiographic images plays an important role in cancer diagnosis, therapy planning and treatment [1, 2]. Many algorithms that solely rely on image-processing techniques have been successfully validated for registration of images of selected organs [1, 3, 4]. However, such algorithms exhibit important deficiencies in capturing large deformations of soft body organs/tissue and skeletal motion associated with registration of whole-body computed tomography (CT) or magnetic resonance (MR) images. For such problems, application of biomechanical models, that utilise the principles of computational mechanics, has been advocated to compute deformations of soft body organs/tissues to register (align) two images [5-7].

Biomechanics-based image registration has been historically dominated by Finite Element Method (FEM). In practice, the 8-noded hexahedral element is a preferable

F. Bello and S. Cotin (Eds.): ISBMS 2014, LNCS 8789, pp. 50–57, 2014.

choice when building the models as it does not exhibit volumetric locking for incompressible/nearly incompressible materials such as soft tissues [8]. However, for abdominal organs and other anatomical structures with complex geometry, spatial discretisation (meshing) using hexahedral elements requires time-consuming manual corrections even if state-of-the-art software specifically designed for generation of meshes of anatomical geometries is applied [9, 10].

To eliminate tedious hexahedral mesh generation, meshless methods of computational mechanics, that use easy-to-generate computational grids in a form of cloud of points, have been proposed in the literature for patient-specific biomechanical models [11, 12]. In this study, we use Meshless Total Lagrangian Explicit Dynamics (MTLED) algorithm [11-13] previously successfully applied in computing the brain deformations for neuro-image registration [11, 14].

Conventionally, the material properties of body organs/tissues are assigned by dividing (segmenting) CT/MR images into non-overlapping constituents with different material properties. Although attempts have been made to automate segmentation algorithms [15], in practice, time-consuming manual correction that relies on analyst skills and somewhat subjective interpretation of the images is often needed. Following the studies [9, 14] on application of patient-specific biomechanical modelling for CT and MR image registration, we eliminate the need for segmentation by assigning the material properties using fuzzy tissue classification. The fuzzy tissue classification relies on the Fuzzy C-Means algorithm to calculate the material properties using fuzzy membership functions for the specified image intensity clusters (corresponding to types of tissue depicted in the image) for each voxel in the image [14, 16]. Such classification is an automated process, and the number of tissue classes (image intensity clusters) is the only parameter that needs to be defined by an analyst.

In this study, we demonstrate feasibility and accuracy of the proposed approach by applying it to register two whole-body CT image-sets (referred to as source and target set) acquires at different time for the same patient. We conduct verification by comparing the deformations, that align the source image-set to the target one, computed using patient-specific models implemented by means of the MTLED algorithm combined with fuzzy tissue classification and traditionally used finite element method that relies on meshing for spatial discretisation. Accuracy of the proposed meshless approach is evaluated by comparing the organ contours in the registered (i.e. source image warped using deformations predicted by our patient-specific meshless model) and target images.

2   Methods

2.1   Analysed Whole-Body CT Image Dataset

The analysed CT image dataset was acquired from the publicly available Slicer Registration Library database, Case #20: Intra-subject whole-body/torso PET-CT (http://www.na-mic.org/Wiki/index.php/Projects:RegistrationLibrary:RegLib_C20b). The dataset consists of two image sets with original resolution of 1mm×1mm×5mm acquired at different time-points for the same patient. The sagittal sections of the two image sets are shown in Fig. 1.

2.2   Meshless Total Lagrangian Explicit Dynamics (MTLED) Algorithm

As the detailed description of the Meshless Total Lagrangian Explicit Dynamics (MTLED) algorithm has been provided in [11-13], only brief summary is given here. Computational efficiency of the MTLED algorithm has been achieved through application of Total Lagrangian (TL) formulation of computational mechanics [17] for updating the calculated variables and Explicit Integration in the time domain combined with mass proportional damping. In the Total Lagrangian formulation, all the calculated variables (such as displacements and strains) are referred to the original configuration of the analysed continuum. The decisive advantage of this formulation is that all derivatives with respect to spatial coordinates can be pre-computed [17, 18], which reduces the number of floating point operations per time step in comparison to Updated Lagrangian (UL) formulation used in vast majority of commercial dynamic finite element and meshless solvers.

The MTLED algorithm relies on modified Galerkin method. The field variable is approximated and interpolated using moving least-squares (MLS) shape functions over the domain geometry discretised using nodes (points) [12]. Following [12], we use regular hexahedral background grid with one integration point per cell for spatial integration. The grid was generated automatically as we do not require the integration cells to conform to the geometry of the anatomical structures depicted in the images. The accuracy of this approach has been previously confirmed by Horton et al. [12, 14].

To illustrate how the domain volume geometry is discretised, Fig. 2 shows the whole-body model discretised using a "cloud" of nodes.

2.3   Loading

Following our previous study [9], we load the meshless model by applying prescribed displacements (essential boundary conditions) on the vertebrae. We select vertebrae as the areas to determine the displacements and prescribe the essential boundary conditions since they can be reliably distinguished from the surrounding tissue in CT images. The displacements between vertebrae in source and target images are calculated using the built-in rigid registration algorithm in the 3D SLICER (http://www.slicer.org/)— an open-source software for visualisation, registration, and segmentation of medical images developed by Artificial Intelligence Laboratory of Massachusetts Institute of Technology and Surgical Planning Laboratory at Brigham and Women's Hospital and Harvard Medical School [19].

2.4   Constitutive Properties and Constitutive Model

Following the approach verified in [9], we use fuzzy tissue classification that utilises the Fuzzy C-Means (FCM) clustering algorithm to assign constitutive properties directly from the images based on the intensity of different tissues of anatomical structures depicted in the images [14, 16].

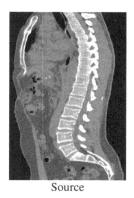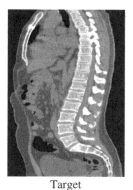

Source   Target

Fig. 1. Sagittal sections of whole-body CT image dataset analysed in this study

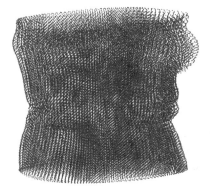

Fig. 2. Whole-body meshless model used in this study. "Cloud" of nodes is used for spatial discretisation.

In the FCM clustering algorithm, each pixel (voxel) in the image is assigned to a number of different tissue types (classes) with different probability. This is done by clustering similar intensity data (pixels) through computation of the membership functions that link the intensity at each pixel with all the specified (i.e. defined by the analyst) cluster centres [14, 16].

There is a vast body of experimental evidence confirming that soft tissues behave like incompressible/nearly incompressible hyperelastic/hyperviscoelastic materials [20, 21]. Therefore, following [20], we used the Neo-Hookean hyperelastic model — the simplest constitutive model that satisfies this requirement. Following [14], the shear modulus G at each integration point of the meshless model was calculated from the FCM-determined membership function u using linear interpolation:

$$G_j = \sum_{k=1}^{C} u_{jk} \times G_k \qquad (1)$$

where G_j is the shear modulus at integration point j, G_k is the shear modulus for a given tissue class k, and C is the number of tissue classes (centres of the intensity clusters in the images). Following [9], we use eight clusters (tissue types) with the shear modulus for each class given in Table 1. As the membership function u is computed from pixel intensity, Equation (1) links the information about image intensity with the mechanical properties of anatomical structures depicted in the image.

Table 1. Shear modulus ($\times 10^3$Pa) of different tissue classes for assigning the constitutive properties at integration points using Eq. (1). Note that voxels with different intensity can depict the same tissue (as for lungs and bones in the image datasets analysed here).

Intensity Cluster Centre	-826	-537	-326	-90	-32	43	274	661
Tissue Classes	Lung	Lung	Lung	Fat	Ligament	Muscle	Bone	Bone
Shear Modulus (kPa)	0.53	0.53	0.53	1.07	3.57	4.05	rigid	rigid

3   Results

3.1   Verification of Meshless Models

We verified the computation results (nodal displacements) predicted by means of the proposed patient-specific meshless model implemented using the MTLED algorithm by comparing them with the results predicted using finite element model created and validated in the study by Li et al. [9]. As shown in Fig. 3, for almost all nodes (over 99.5%), the differences between the displacements predicted using the two models are less than 1 mm and the maximum difference is 2.8mm. As these differences are appreciably less than the resolution (1mm×1mm×5mm) of whole-body CT image dataset analysed here, they can be safely treated as negligible.

3.2   Evaluation of Registration Accuracy

Following Li et al. [9] and Mostayed et al. [22], we evaluated the registration accuracy by comparing the edges/contours of the organs in registered (i.e. source image warped using the deformations predicted by our patient-specific meshless model) and target images. As the lung is a large organ that can be reliably distinguished, we qualitatively evaluate the registration accuracy by comparing the contours of the lung extracted from the registered and target images. It can be seen (Fig. 4) that the lung contour extracted from the registered image are very close to those extracted from the target image. The distance between the two contours (registration error) is less than two times of image voxel size of the source image. As the registration accuracy is limited by the image resolution, the edge/contour pair having a distance less than two times of voxel size of the original source image can be regarded as successfully registered.

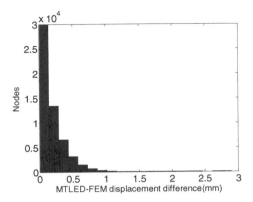

Fig. 3. Verification of the meshless model with fuzzy tissue classification used here for whole-body image registration. Differences between the nodal displacements predicted using the meshless model created in this study and previously validated [9] finite element model.

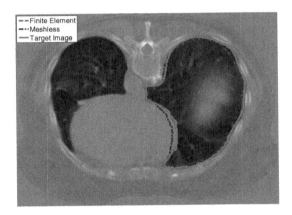

Fig. 4. Evaluation of registration accuracy. The dotted line and dashed line (they are nearly overlapping) represent lung contours extracted from images registered using deformations predicted by means of the meshless model used in this study and previously validated finite element model [9], respectively. The solid line is the lung contour extracted from the target image.

4   Discussion and Conclusions

In this study, we demonstrate feasibility of patient-specific computation of organ/tissue deformation for registration of whole-body CT images using meshless discretisation combined with fuzzy tissue classification. Unlike traditionally used finite element method, meshless discretisation (we used Meshless Total Lagrangian Explicit Dynamics MTLED algorithm with hexahedral background integration grid) facilitates automated and rapid generation of computational grid. Fuzzy tissue classification eliminates the need for tedious image segmentation to subdivide the image into non-overlapping regions that correspond to different tissue types. Instead, the material properties are assigned at the integration points directly from the image. Fuzzy C-Means algorithm is adapted here to calculate the material properties using the fuzzy membership functions.

For the whole-body CT image dataset analysed in this study, the displacements predicted using the proposed meshless approach were very close to those previously obtained by Li et al. [9] using the validated finite element model. The organ contours in the registered (i.e. source image warped using deformations predicted by our patient-specific meshless model) image were close to the contours in the target image. The distance between the organ contours in the registered and target images was within the commonly used accuracy threshold of two times the voxel size.

Acknowledgments. The first author is a recipient of the SIRF scholarship and acknowledges the financial support of the University of Western Australia. The support of Australian Research Council (Discovery Grant DP120100402) is gratefully acknowledged. The whole-body CT image dataset analysed in this study was acquired from publically available Registration Library by National Alliance for Medical Image Computing (NA-MIC). NA-MIC is a national research centre supported by grant U54 EB005149 from the NIBIB NIH HHS Roadmap for Medical Research Program. The authors acknowledge contribution of Dr Guiyong Zhang (at present Dalian University of Technology, work done during appointment at The University of Western Australia) to development of the meshless code used in this study.

References

1. Warfield, S.K., Haker, S.J., Talos, I.F., Kemper, C.A., Weisenfeld, N., Mewes, A.U.J., Goldberg-Zimring, D., Zou, K.H., Westin, C.F., Wells, W.M., Tempany, C.M.C., Golby, A., Black, P.M., Jolesz, F.A., Kikinis, R.: Capturing intraoperative deformations: research experience at Brigham and Women's Hospital. Med. Image Anal. 9(2), 145–162 (2005)
2. Black, P.M., Moriarty, T., Alexander, E., Stieg, P., Woodard, E.J., Gleason, P.L., Martin, C.H., Kikinis, R., Schwartz, R.B., Jolesz, F.A.: Development and implementation of intra-operative magnetic resonance imaging and its neurosurgical applications. Neurosurgery 41(4), 831–842 (1997)
3. Mattes, D., Haynor, D.R., Vesselle, H., Lewellen, T.K., Eubank, W.: PET-CT image registration in the chest using free-form deformations. IEEE T. Med. Imaging 22(1), 120–128 (2003)
4. Goerres, G.W., Kamel, E., Heidelberg, T.N.H., Schwitter, M.R., Burger, C., von Schulthess, G.K.: PET-CT image co-registration in the thorax: influence of respiration. Eur. J. Nucl. Med. Mol. I 29(3), 351–360 (2002)
5. Hagemann, A., Rohr, K., Stiehl, H.S., Spetzger, U., Gilsbach, J.M.: Biomechanical modeling of the human head for physically based, nonrigid image registration. IEEE T. Med. Imaging 18(10), 875–884 (1999)
6. Chanthasopeephan, T., Desai, J.P., Lau, A.C.W.: Modeling soft-tissue deformation prior to cutting for surgical simulation: Finite element analysis and study of cutting parameters. IEEE T. Bio-Med. Eng. 54(3), 349–359 (2007)
7. Otoole, R.V., Jaramaz, B., Digioia, A.M., Visnic, C.D., Reid, R.H.: Biomechanics for Preoperative Planning and Surgical Simulations in Orthopedics. Computers in Biology and Medicine 25(2), 183–191 (1995)

8. Irving, G., Teran, J., Fedkiw, R.: Tetrahedral and hexahedral invertible finite elements. Graph Models 68(2), 66–89 (2006)
9. Li, M., Wittek, A., Joldes, G., Zhang, G., Dong, F., Kikinis, R., Miller, K.: Whole-Body Image Registration Using Patient-Specific Non-Linear Finite Element Model. In: Doyle, B.J., Miller, K., Wittek, A., Nielsen, P.M.F. (eds.) Computational Biomechanics for Medicine: Fundamental Science and Patient-Specific Application, pp. 21–30. Springer, New York (2013)
10. Wittek, A., Joldes, G., Couton, M., Warfield, S.K., Miller, K.: Patient-specific non-linear finite element modelling for predicting soft organ deformation in real-time; Application to non-rigid neuroimage registration. Prog. Biophys. Mol. Bio. 103(2-3), 292–303 (2010)
11. Miller, K., Horton, A., Joldes, G.R., Wittek, A.: Beyond finite elements: A comprehensive, patient-specific neurosurgical simulation utilizing a meshless method. J. Biomech. 45(15), 2698–2701 (2012)
12. Horton, A., Wittek, A., Joldes, G.R., Miller, K.: A meshless Total Lagrangian explicit dynamics algorithm for surgical simulation. Int. J. Numer. Meth. Bio. 26(8), 977–998 (2010)
13. Zhang, G.Y., Wittek, A., Joldes, G.R., Jin, X., Miller, K.: A three-dimensional nonlinear meshfree algorithm for simulating mechanical responses of soft tissue. Engineering Analysis with Boundary Elements 42, 60–66 (2014)
14. Zhang, J.Y., Joldes, G.R., Wittek, A., Miller, K.: Patient-specific computational biomechanics of the brain without segmentation and meshing. Int. J. Numer. Meth. Bio. 29(2), 293–308 (2013)
15. Balafar, M.A., Ramli, A.R., Saripan, M.I., Mashohor, S.: Review of brain MRI image segmentation methods. Artif. Intell. Rev. 33(3), 261–274 (2010)
16. Bezdek, J.C., Ehrlich, R., Full, W.: Fcm - the Fuzzy C-Means Clustering-Algorithm. Comput. Geosci. 10(2-3), 191–203 (1984)
17. Miller, K., Joldes, G., Lance, D., Wittek, A.: Total Lagrangian explicit dynamics finite element algorithm for computing soft tissue deformation. Commun. Numer. Meth. En. 23(2), 121–134 (2007)
18. Joldes, G.R., Wittek, A., Miller, K.: Suite of finite element algorithms for accurate computation of soft tissue deformation for surgical simulation. Med. Image Anal. 13(6), 912–919 (2009)
19. Fedorov, A., Beichel, R., Kalpathy-Cramer, J., Finet, J., Fillion-Robin, J.C., Pujol, S., Bauer, C., Jennings, D., Fennessy, F., Sonka, M., Buatti, J., Aylward, S., Miller, J.V., Pieper, S., Kikinis, R.: 3D Slicer as an image computing platform for the Quantitative Imaging Network. Magn. Reson. Imaging 30(9), 1323–1341 (2012)
20. Miller, K., Wittek, A., Joldes, G.: Biomechanical Modeling of the Brain for Computer-Assisted Neurosurgery. In: Miller, K. (ed.) Biomechanics of the Brain, pp. 111–136. Springer, New York (2011)
21. Fung, Y.C.: Mechanical Properties of Living Tissues. Springer, New York (1993) ISBN 978-1-4757-2257-4
22. Mostayed, A., Garlapati, R.R., Joldes, G.R., Wittek, A., Roy, A., Kikinis, R., Warfield, S.K., Miller, K.: Biomechanical Model as a Registration Tool for Image-Guided Neurosurgery: Evaluation Against BSpline Registration. Ann. Biomed. Eng. 41(11), 2409–2425 (2013)

Automatic Alignment of Pre and Intraoperative Data Using Anatomical Landmarks for Augmented Laparoscopic Liver Surgery

Rosalie Plantefève[1,2], Nazim Haouchine[2], Jean-Pierre Radoux[1], and Stéphane Cotin[2,3]

[1] MEDIC@, ALTRAN, France
[2] Shacra Team, INRIA, France
[3] IHU Strasbourg, France

Abstract. Each year in Europe 50,000 new liver cancer cases are diagnosed for which hepatic surgery combined to chemotherapy is the most common treatment. In particular the number of laparoscopic liver surgeries has increased significantly over the past years. This type of minimally invasive procedure which presents many benefits for the patient is challenging for the surgeons due to the limited field of view. Recently new augmented reality techniques which merge preoperative data and intraoperative images and permit to visualize internal structures have been proposed to help surgeons during this type of surgery. One of the difficulties is to align preoperative data with the intraoperative images. We propose in this paper a semi-automatic approach for solving the ill-posed problem of initial alignment for Augmented Reality systems during liver surgery. Our registration method relies on anatomical landmarks extracted from both the laparoscopic images and three-dimensional model, using an image-based soft-tissue reconstruction technique and an atlas-based approach, respectively. The registration evolves automatically from a quasi-rigid to a non-rigid registration. Furthermore, the surface-driven deformation is induced in the volume via a patient specific biomechanical model. The experiments conducted on both synthetic and *in vivo* data show promising results with a registration error of 2 mm when dealing with a visible surface of 30% of the whole liver.

1 Introduction

Context: Minimally invasive surgery is regarded as one of the major advances in surgery of last decades considering the benefits for patient in term of time recovery and reduced pain. Yet, one problem remains: the surgeon still needs to register mentally the preoperative data onto the laparoscopic view to determine the tumours and vessels location. The introduction of optics in the clinical routines have motivated several research groups to investigate the benefits of Augmented Reality to help surgeons during the procedure [1]. Indeed, Augmented Reality has the potential to provide an enriched visual feedback to the surgeon by the fusion of intraoperative images and three-dimensional preoperative data such as tumours and vascular network.

F. Bello and S. Cotin (Eds.): ISBMS 2014, LNCS 8789, pp. 58–66, 2014.

Previous Work: In this context several methods have been proposed to solve the numerous challenges that present the abdominal cavity, ranging from occluded regions due to instrument [2], organ tracking disturbance due to blur or smoke [3], and unpredictable deformations produced by heart beating, breathing or manipulation with instruments [4]. However, despite these numerous techniques, very few works have investigated the initial alignment between the laparoscopic images and the three-dimensional model. Indeed, the initialisation is often assumed to be done manually [2,4,5], which can be a major source of error given the significant deformations that the organs may undergo during the surgery. In addition, since gas is insufflated (pneumoperitoneum) to increase the working space [6], the preoperative data may no longer correspond to the intraoperative image. In order to optimize the trocars placement on the skin, in [6] it is proposed to simulate the pneumoperitoneum using only a preoperative model and patient-specific biomechanical parameters. The evaluation of their method on real data highlights its potential usability for Augmented Reality. Based on this work, Oktay *et al.* [7] propose to exploit the intraoperative data acquired from CT-scans after insufflation as an additional constraint to drive the simulation. Although this method provides accurate registration, it relies on intraoperative scans. Another intraoperative registration method is presented in [8]. The authors use an intraoperative ultrasound probe to register the vessel tree on a three dimentional liver model. However, since our aim is to register the laparoscopic data recorded by a camera, such information is not available. Clements *et al.* [9] introduced an ICP-based approach to aid in the initial pose estimation. Using salient anatomical features, identifiable in both the preoperative images and intraoperative liver surface data, this method is able to reach a reasonable solution, but is restrained to rigid transformations. To handle deformations, biomechanical models are often used to regularize the registration methods. For instance, in [10] a registration of intraoperative MR brain images is proposed where the model is based on linear elasticity and the finite element method. Moreover the method proposed by Clements relies on manual estimation of anatomical structures position which is too cumbersome to be done in a clinical context. A solution could be to use anatomical atlases to gather information about features positions. In [11] the authors use a histological atlas containing detailed anatomical information to target small brain structure in a pre operative MRI for neurosurgery planning.

2   Methods

The overall computational flow of our method involves two main steps: detection of anatomical landmarks on both laparoscopic images and preoperative data, and a non-rigid feature-based registration similar to the Iterative Closest Point (ICP).

2.1   Pre and Intraoperative Data Computation

We define a set of landmarks that are chosen to be visible on stereoscopic images and in at least one preoperative imaging technique (see Fig. 1). These landmarks will be labelled on one frame of the endoscopic data and on the three-dimensional model.

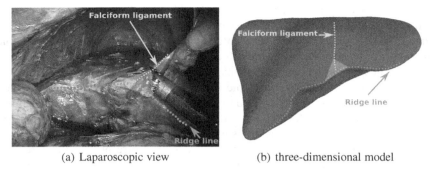

(a) Laparoscopic view   (b) three-dimensional model

Fig. 1. Our registration method must acurately register the three-dimensional model onto the laparoscopic view. The ridge line and a section of the falciform ligament are used as anatomical features to guide the registration process.

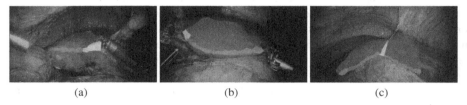

(a)   (b)   (c)

Fig. 2. Detection of anatomical landmarks on laparoscopic images: The liver surface (in green), the ridge line (in orange) and the falciform ligament (in yellow) are extracted from the laproscopic image using stereo matching method to obtain a 3D labelled point cloud.

Stereo Endoscopic Scene Estimation: On one of the stereo endoscopic images, the falciform ligament, the ridge line and the liver surface are defined manually as anatomical landmarks of the liver. The operator selects (e.g. by clicking) some points on the border of the regions of interest to form closed areas. This selection takes only a few seconds. Once the targeted regions are selected, we build for each of them a three-dimensional point cloud using the method described in [4] that relies on a sparse stereo matching technique to recover the liver surface shape. These point clouds are labelled according to their corresponding regions as illustrated in Fig. 2.

Preoperative Anatomical Landmarks Detection: The three-dimensional model is obtained form the CT images via a standard semi-automatic segmentation process described in [12]. The resulting mesh is then augmented with the same anatomical landmarks as the ones detected in the stereoscopic images. For the liver one landmark is automatically detected on the three-dimensional model: the ridge line. This landmark can be easily identified in most human livers. The automatic detection of the ridge line on the three-dimensional model of the liver is performed as follows: the edges separating two triangles with sufficiently different normals are selected as seeds. The definition of the threshold is based on statistics on the normals differences for the whole mesh.

Then, the ridge line is extended from the seed edges; if no extension is found the seeds are removed. Iteratively the ridge line is reconstructed. The other landmarks (only the falciform ligament for the liver) are transferred via a statistical atlas of the liver surface containing the landmarks location [13]. This atlas is constructed with a set of pre operative medical images of the organ of interest. The imaging modality should be chosen such that the selected anatomical landmarks are visible. For each image an expert perform a manual segmentation of the organ and anatomical landmarks. Then all the segmentations are aligned in a common reference frame and the mean shape and standard deviations are computed. This atlas is finally registered on the three-dimensional model.

2.2 Feature-Based Registration

Our approach relies on a non-rigid registration which computes the elastic transformation that maximizes the shape similarity between the source and target configurations. One ICP is computed per labelled area and they are all solved simultaneously.

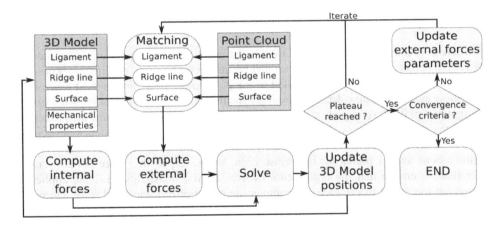

Fig. 3. Main steps of the ICP-like registration method

Matching: The target model is the sets of points extracted from the stereoscopic images and the source model is the organ three-dimensional model. Three matchings are achieved simultaneously: one for each labelled area. For each area the points of the target model are projected on their corresponding area on the source model surface as shown in Fig. 3. We do not project the source model points on the target since the target only corresponds to a part of the source model. The target points are projected onto the triangles of the source model surface. A matching pair $\{p_t, p_s\}$ consisting of the target point and its closest projection is then defined. The outliers are pruned with relative distance and normal thresholds. The relative distance thresholds keep the pairs points for which $\|\overrightarrow{p_t p_s}\| < d_t \max(\{\|\overrightarrow{p_t p_s}\|\}_{all\,pairs})$, $d_t \in [0; 1]$ where $\|\cdot\|$ is the Euclidean norm. The normal thresholds prune the pairs for which their normal dot product $\widehat{n}_{p_t} \cdot \widehat{n}_{p_s} < n_t$,

$n_t \in [0; 1]$. These thresholds must be set according to the deformation characteristics. The normal should be smaller and the distance threshold larger for larger deformations. The ICP algorithm aims at reducing the global distance between the $\{p_t, p_s\}$ pairs.

Model of Deformation: A biomechanical model is associated to the augmented three-dimensional model. As the three-dimensional model and the points extracted from the stereoscopic images are two representations of the same liver, they share the same biomechanical behaviour. Different constitutive laws can be used to model the liver such as hyperelastic formulation [14]. For computational efficiency we rely on a co-rotational formulation as proposed by [15]. A fast finite element approach is used to compute the deformation of the organ. This computation can be performed very efficiently thus making it very suitable for the ICP algorithm. Regarding the parameters of the model, the Young's modulus can be patient specific (e.g. obtained by elastography) or is set to the average value found in the literature [16]. The Poisson ratio is set to 0.4 as the liver is nearly incompressible (fluid exchanges allow for some volume variation).

Definition of Constraints: The constraints are imposed on the system with a penalty force:

$$\mathbf{f} = \sum_i^m \mathbf{f}_i, \quad (1) \qquad\qquad \text{with} \quad \mathbf{f}_i = \sum_j^{p_i} f_i \arctan(k\mathbf{x}_{ij}) \quad (2)$$

where m is the number of labelled area, p_i the number of corresponding pairs in the area i, f_i a penalty factor, k a scale factor and $\mathbf{x}_{ij}$ the vector between the j-th target point and its projection. The k value is related to the point cloud noise and is used to avoid overfitting as a lower value decrease the force intensity for small distances. When registering non-rigidly the complete surface model w.r.t. the reconstructed part, the process may suffer from inaccuracy due to the limited knowledge of correspondence between the two. Therefore, the labelled areas are penalized differently to use the anatomical landmarks as anchor points for the registration. Moreover, all the f_i are not constant over time to ensure that the registration evolves from a quasi-rigid (i.e. the applied forces are too small to produce a significant deformation) to non-rigid state. During the registration we detect when the mechanical system has reached a plateau and then update the f_i values. In case of liver, the falciform ligament is the most reliable anatomical feature. It also corresponds to a very small area and therefore does not contribute much to the deformation. For that reason, f_i corresponding to this landmark is constant during the registration process. The other f_i are of the form $\frac{2f_{i_{max}}(\frac{n}{v_i})^2}{(\frac{n}{v_i})^2+1}$ for $n < v_i$ (i.e. they are set to 0 at the beginning of the registration) and $f_{i_{max}}$ for $n > v_i$ where n is the plateau index and v_i a speed coefficient. v_i and $f_{i_{max}}$ must be set so that the objective function is more convex in the early stage of registration and that the minimum become more precise over time. Therefore, the v_i and $f_{i_{max}}$ of small area must be smaller than those corresponding to the large area. According to our tests, the algorithm shows low sensitivity w.r.t. the parameters. The registration is stopped when $n = \max(v_i) + 1$.

Registration Method: At each simulation step the following equation is solved

$$\mathbf{K}(\mathbf{u})\mathbf{u} = \mathbf{f} \qquad\qquad\qquad (3)$$

where $\mathbf{u}$ is the displacement vector, $\mathbf{K}$ the non-linear stiffness provided by the FE formulation and $\mathbf{f} = \frac{dE(\mathbf{u})}{d\mathbf{u}}$ is the vector of constraint forces. The pairing is then recomputed generating a new force. Finally, during the simulation the anatomical landmarks provide a coarse registration that improves the robustness of the ICP method, whereas the biomechanical model plays a role of regularization and allows for an accurate solution of local deformations.

3   Results

This section presents the results obtained using our method. We conduct experiments on two different type of data: computer-generated data and liver data during *in vivo* surgical procedure. For convergence an average of 3700 iterations is needed. Each iteration takes 700 ms on average and the solver represent around 20% of a time step. In order to insure the success of the method, the anatomical features must be at least partially visible.

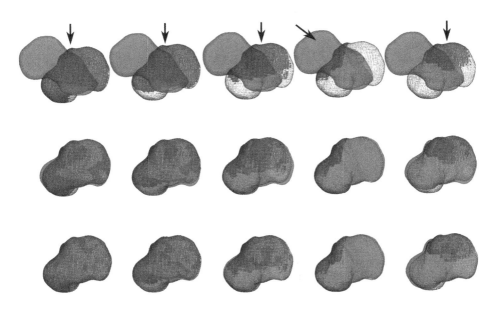

Fig. 4. Registration results on synthetic data. The source model is in transparent red, the target surface used for the registration is in blue and the whole target surface in green wireframe. The first row shows the initial pose, the second row the result after the quasi-rigid registration and the last row the final result. Each column shows the results for different target surface areas. The target surface of the columns from left to right correspond to 50%, 40%, 30%, 20% and 10% of the whole surface. The black arrows indicate the virtual camera angle.

3.1   Evaluation with *in silico* Data

We evaluated our method by registering a three-dimensional model of the liver segmented from CT-scans (source) on the same model that had undergone both a rigid and
non-rigid transformations (target). In order to obtain the target liver, we apply a pressure
on the surface of the model to simulate two different deformations possibly induced by
the pneumoperitoneum, and a random rigid transformation (cf Fig. 4). In real situations
only partial surface information is acquired by the laparoscopic camera. To evaluate the
amount of information needed to achieve an accurate registration faces of the deformed
model are deleted to keep only a portion of its surface from 50% to 10%.

The results presented in Fig. 5 show that even with only 10% of visible surface and a
poor initial alignment our method is able to perform the registration with a mean error
lower than 4 mm. The deformation 1 was more important on the liver face opposed
to the camera view. This fact could explain the lower quality of the results. The plot
also suggests that for a small deformation the non-rigid part of the method does not
introduce any improvement when compared to the quasi-rigid part as soon as 20% or
less of the surface is visible. In these two cases the visible area underwent a small
deformation comparatively to the other cases. However, the mean Hausdorff distance
calculated between the registered three-dimensional model and the visible part does not
exceed 0.3 mm.

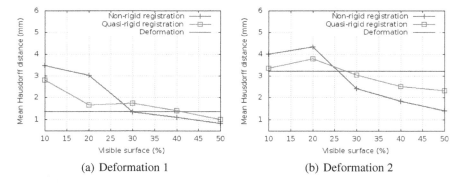

(a) Deformation 1   (b) Deformation 2

Fig. 5. Registration results on synthetic data showing the error w.r.t. the variation in visible surface
area for deformation 1 and 2. Liver Young modulus: $YM = 50kPa$, $f_{liga_{max}} = YM/10$, $f_{ridge_{max}} = YM/3$, $v_{ridge} = 2$, $f_{surf_{max}} = 2YM$, $v_{surf} = 25$, $d_t = 0.95$, $n_t = 0$ and $k = 2\pi.10^3$ (90% of the
force intensity at 10^{-3} mm). The blue lines represent the mean Hausdorff distance between the
non deformed source model and the same model after deformation.

3.2   Experiments with *in vivo* Data

To assess our approach in a real surgical environment (specular lights, instrument oc-
clusions ...) and to evaluate the ability of our non-rigid registration to estimate the
initial pose, we tested our method on *in vivo* laparoscopic images of a human liver.
The results illustrated in Fig. 6 report a visually correct initial non-rigid registration of

the liver model on the laparoscopic image with only a partial and noisy three-dimensional surface reconstruction. For this simulation we used the same parameter values as for the *in silico* data (see the caption of Fig. 5) except for $k = 2\pi.10^1$ (90% of the force intensity at 10^{-1} mm). The mean hausdorff distance between the point clouds and the three-dimensional model is 0.5 mm and 0.6 mm for the cases (a) and (b) of Fig. 6 respectively.

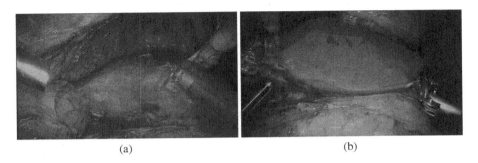

(a)   (b)

Fig. 6. Registration results on *in vivo* data on two different views of a human liver. The registered mesh is shown in red while the partial reconstructed patch is depicted in blue.

4   Conclusion

In this paper, we proposed a semi-automatic method for initial alignment of an organ preoperative three-dimensional model onto an endoscopic view of the same organ. This method relies on a feature-based registration framework which evolves from a quasi-rigid state to a non rigid state. The anatomical features used to guide the registration are selected manually on the endoscopic view and are transferred via an atlas to the three-dimensional model. We evaluated the method using ten synthetic data and two real endoscopic views. The results show that the method deforms accurately the three-dimensional model when the visible surface corresponds to 30% or more of the entire organ surface. It was also demonstrated that the registration works with real endoscopic data. In the future, we will add information about the organ silhouette. We also plan to limit the constraint forces applied during the registration according to real forces imposed on the organ boundary during the intervention.

Acknowledgement. The authors would like to thank the anonymous reviewers for their valuable comments and suggestions and Igor Peterlik for his thorough re-reading of this article.

References

1. Nicolau, S., Soler, L., Mutter, D., Marescaux, J.: Augmented reality in laparoscopic surgical oncology. Surgical Oncology 20(3), 189–201 (2011)
2. Puerto-Souza, G., Mariottini, G.: Toward long-term and accurate augmented-reality display for minimally-invasive surgery. In: ICRA 2013, pp. 5384–5389 (2013)

3. Yip, M.C., Lowe, D.G., Salcudean, S.E., Rohling, R., Nguan, C.Y.: Tissue tracking and registration for image-guided surgery. IEEE Trans. Med. Imaging 31(11), 2169–2182 (2012)
4. Haouchine, N., et al.: Image-guided simulation of heterogeneous tissue deformation for augmented reality during hepatic surgery. In: ISMAR (2013)
5. Su, L.M., et al.: Augmented reality during robot-assisted laparoscopic partial nephrectomy: Toward real-time 3d-ct to stereoscopic video registration. Urology 73(4), 896–900 (2009)
6. Bano, J., et al.: Simulation of pneumoperitoneum for laparoscopic surgery planning. In: Ayache, N., Delingette, H., Golland, P., Mori, K. (eds.) MICCAI 2012, Part I. LNCS, vol. 7510, pp. 91–98. Springer, Heidelberg (2012)
7. Oktay, O., Zhang, L., Mansi, T., Mountney, P., Mewes, P., Nicolau, S., Soler, L., Chefd'hotel, C.: Biomechanically driven registration of pre- to intra-operative 3D images for laparoscopic surgery. In: Mori, K., Sakuma, I., Sato, Y., Barillot, C., Navab, N. (eds.) MICCAI 2013, Part II. LNCS, vol. 8150, pp. 1–9. Springer, Heidelberg (2013)
8. Dagon, B., Baur, C., Bettschart, V.: A framework for intraoperative update of 3d deformable models in liver surgery. In: 30th Annual International Conference of the IEEE Engineering in Medicine and Biology Society, EMBS 2008, pp. 3235–3238. IEEE (2008)
9. Clements, L.W., et al.: Robust surface registration using salient anatomical features for image-guided liver surgery: Algorithm and validation. Medical Physics 35(6) (2008)
10. Ferrant, M., et al.: Registration of 3-d intraoperative mr images of the brain using a finite-element biomechanical model. IEEE Trans. on Medical Imaging 20(12), 1384–1397 (2001)
11. Bardinet, E., et al.: A three-dimensional histological atlas of the human basal ganglia. atlas deformation strategy and evaluation in deep brain stimulation for parkinson disease. J. of Neurosurgery 110(2) (2009)
12. Yushkevich, P.A., Piven, J., Cody Hazlett, H., Gimpel Smith, R., Ho, S., Gee, J.C., Gerig, G.: User-guided 3D active contour segmentation of anatomical structures: Significantly improved efficiency and reliability. Neuroimage 31(3), 1116–1128 (2006)
13. Plantefève, R., Peterlik, I., Courtecuisse, H., Trivisonne, R., Radoux, J.-P., Cotin, S.: Atlas-based transfer of boundary conditions for biomechanical simulation. In: Golland, P., Hata, N., Barillot, C., Hornegger, J., Howe, R. (eds.) MICCAI 2014, Part II. LNCS, vol. 8674, pp. 33–40. Springer, Heidelberg (2014)
14. Marchesseau, S., Heimann, T., Chatelin, S., Willinger, R., Delingette, H.: Multiplicative jacobian energy decomposition method for fast porous visco-hyperelastic soft tissue model. In: Jiang, T., Navab, N., Pluim, J.P.W., Viergever, M.A. (eds.) MICCAI 2010, Part I. LNCS, vol. 6361, pp. 235–242. Springer, Heidelberg (2010)
15. Felippa, C., Haugen, B.: A unified formulation of small-strain corotational finite elements: I. theory. Comput. Meth. Appl. Mech. Eng. 194(21) (2005)
16. Yeh, W.C., et al.: Elastic modulus measurements of human liver and correlation with pathology. Ultrasound in Medicine & Biology (2002)

Using a Biomechanical Model
for Tongue Tracking in Ultrasound Images

Matthieu Loosvelt, Pierre-Frédéric Villard, and Marie-Odile Berger

LORIA/CNRS, Université de Lorraine, INRIA, France
{firstname.lastname}@loria.fr

Abstract. We propose in this paper a new method for tongue tracking in ultrasound images which is based on a biomechanical model of the tongue. The deformation is guided both by points tracked at the surface of the tongue and by inner points of the tongue. Possible uncertainties on the tracked points are handled by this algorithm. Experiments prove that the method is efficient even in case of abrupt movements.

Keywords: tracking, biomechanical model, tongue, ultrasound.

1 Introduction

The shape and dynamics of the tongue during speech provide valuable information on the speech production system. Currently, ultrasound imagery is widely recognized as the best way to acquire information on tongue movements at a fast frame rate [2,14]. As speech applications require the exploitation of rather long speech recording runs, automatic extraction procedures are required. Automatic tracking in US images is known to be very difficult due to low signal-to-noise ratio, high speckle noise, acoustic shadowing, mirroring... This often results in missing parts in the observed contour, especially for fast tongue movements. The interested reader may refer to [11] for an overview of segmentation techniques dedicated to US data. To cope with these problems, there have been many attempts to incorporate prior information on the tongue shape variations to guide the detection. These solutions can be classified into two broad categories: Deformable model-based techniques proposed various extensions of the *Snake* model where contours are guided towards points with high intensity gradients and are submitted to spatial and sometimes temporal regularization constraints to deal with noisy images [10,1,15]. On the other hand, learning based techniques make use of manual delineations or of markers glued on the tongue to predict the tongue position in images [12,4]. However such methods require to manually extract the tongue on large data sets: in [12], 700 X-ray images were considered for training.

In all these tracking methods, the main difficulty is to incorporate only plausible tongue shape deformations within the tracking framework. Regularization techniques bring too general smoothing constraints to be efficient for fast tongue motions whereas learning-based methods need tedious delineation tasks. We propose in this paper a tracking framework based on a biomechanical model which

F. Bello and S. Cotin (Eds.): ISBMS 2014, LNCS 8789, pp. 67–75, 2014.
© Springer International Publishing Switzerland 2014

appropriately characterizes the elastic properties of the tongue and allows tracking even for fast motions.

2   Related Works

The use of mechanical models to improve tracking performance is a subject of increasing interest in the computer vision community [16,8]. The main interest of such models is to improve tracking on parts of the objects with unobserved features. Constraints deduced from the elastic properties of the material generally provide better prediction than general regularization constraints. However, tracking based on mechanics requires to choose the appropriate mechanical models along with its elastic parameters. Fortunately, the mechanical properties of the tongue have been extensively studied [6] and it has been proven that a hyperelastic model is quite appropriate. We thus investigate in this paper the use of a biomechanical model for tongue tracking. To the best of our knowledge, this approach has not been considered in the challenging context of tongue tracking. Past methods are in fact only dedicated to track the tongue contour whereas we aim to track the tongue volume, taking into account keypoints inside the tongue in addition to contour points. Another contribution of the paper is to properly handle uncertain or false point matchings which are common in such noisy images.

The way we extract correspondences over time is described in section 3. The mechanical model which takes into account these features is described in section 4. Results and comparison with existing methods are highlighted in section 5.

3   Extracting Point Correspondences over Time

We describe in this section how we extract point correspondences over time. The procedure starts with the delineation of a closed contour which contains the upper contour of the tongue and defines the physical region which has to be tracked (see Fig. 1.a). First, the Harris detector is used to find the most prominent corners in the first image of the sequence within this region. These corners are sorted by their quality measure in the descending order. The first 100 features are displayed in Fig. 1.a. The displacement of these points in each subsequent frame is computed thanks to the Farneback's optic flow algorithm [5]. In this approach, the displacement field is supposed to be only slowly varying so that the displacement is computed over a neighborhood of each point with a least square criteria. In addition, we compute the covariance on the estimated displacement at each point. This can be done easily since the estimated covariance on the least squares estimate $\hat{x}$ of the problem $Ax = b$ is given by $\Lambda_x = (A^t A)^{-1}$. Computing this covariance is important since points inside the tongue have in general non ambiguous correspondences and thus a relatively small covariance matrix. On the other hand, since the curvature of the tongue is rather small with homogeneous intensities on both sides of the contours, the points extracted on the tongue contour have a tangential inacurracy in the contour direction and thus an elongated covariance matrix (see Fig. 1.b). The interest of computing the covariance matrix on the estimated displacement is twofold. First, it allows us to remove features which are not reliable according to the eigenvalues of the matrix (in practice, matrices with two large eigenvalues of the same order are removed). Second, these uncertainties are integrated within the biomechanical model (Fig. 2). In order to have a relatively small number of evenly distributed features, a minimal distance d between selected features is imposed (15-30 pixels in practice). Features are chosen in increasing order of their covariance matrix. This explains why features with a relatively large covariance matrix are kept in regions where few features were initially detected.

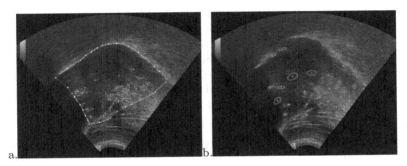

Fig. 1. (a) The tongue region delineated on the first image of the sequence and the extracted features. (b) Covariance computed on the selected features with $d = 20$ pixels on another image of the sequence.

4   The Biomechanical Model

Biomechanical models consist in computing 3D object deformations using physical laws. In order to have accurate results we choose to model our problem using the continuum mechanics formulation (conservation laws, matter continuity, etc.). Various numerical techniques exist to solve this set of equations. We use the finite element technique (FEM) to compute the equations of continuous mechanics. We first define the discrete geometry for dividing the complex problem into small elements. The problem is a mechanical system expressed in ordinary differential equations. We then establish the boundary conditions that the system must satisfy. One of the system equations is the constitutive law that depends on the material properties. Finally we point out how we solve the system.

The Geometry
For speech applications, modeling the tongue in the sagittal plane is sufficient. We use a 2D mechanical model with a 2D geometry where all the motions and deformations occur only in a plane. We thus adopt a quasi-plane strain model, which precisely consists in only focusing on deformations in 2D.

Contours are extracted on the ultrasound image by delineating the upper tongue contour and closing the shape with anatomical landmarks (Fig. 1.a). This 2D polyline is meshed using the "meshAdapt" algorithm [7] while specifying a minimum and maximum edge length of respectively 0.5 and 2 mm. We also impose the points where those displacements have been tracked (see §3) to belong to the mesh. See the resulting mesh on Fig. 2.

The Boundary Conditions and the Constitutive Law

Defining boundary conditions in biomechanical problems is not an easy task as they often result from a complex mechanism of muscle contractions and relaxations. Tongue material does not follow a basic elastic law. As in [6], we chose to model its behavior with a hyper-elastic constitutive equation. We used a first order Mooney-Rivlin law converted from a Yeoh strain-energy function with the elasticity parameter values found in [6]. The model is guided by image-based external forces described in the next section. It must be pointed out that a more sophisticated model of the tongue has been proposed in [3] where the tongue is described by 11 groups of muscle geometries. Besides the fact that the anatomy of the speaker is required, such a model is not appropriate for tracking since it would require too many feature points to be controlled.

The Forces Monitoring the Model

The strategy is as follows: instead of modeling the muscle actions we simulate the effects of some key vertices in the tongue as if these points were manipulated by inserted pins. Forces are then dynamically applied to each mesh node n corresponding to the tracked points. The force is linearly transmitted to the neighbors nodes with a kernel factor providing the number of affected neighbors. We chose this factor to be equal to five because it is the best trade-off between shape regularization and mechanical result accuracy.

Each node n displacement information is converted into a spring force $\mathbf{F}_t^n$ dependent on time t with eq. (1). k is a material constant that affects the convergence rapidity, $\mathbf{P}_F^n$ is the final position of a point n given by Farneback's method and $\mathbf{P}_t^n$ is the node position at a given time t after a biomechanical simulation. On Fig. 2 $\mathbf{P}_F^n$ are the green squares and $\mathbf{P}_t^n$ are the blue ones.

$$\mathbf{F}_t^n = k(\mathbf{P}_F^n - \mathbf{P}_t^n) \tag{1}$$

The simulation stops when the $\mathbf{P}_t^n$ gets reasonably closed to the $\mathbf{P}_F^n$.

Handling Possible Erroneous Features

Due to the high level of noise in US images, some features may be tracked with a rather large inaccuracy which is estimated by our algorithm. However, those features are kept and used to guide the mechanical model since they introduce information in areas where few features are available. To take into account the covariance matrix Λ_F^n on the features P_F^n, a Mahalanobis distance is used to compute the potential energy of the spring systems, which gives rise to the modified forces:

$$\mathbf{F}_t^n = k(\Lambda_F^n)^{-1}(\mathbf{P}_F^n - \mathbf{P}_t^n) \tag{2}$$

The System Resolution
Ordinary differential equations are solved using the finite-element method. They are integrated through an implicit backward Euler method [13] that is reasonably stable for small time step. It is not unconditionally stable because it uses a semi-implicit integration rather than a fully implicit integration. Tongue shape undergoes significant changes during speaking. Large deformation are then taken into account by using the Inversible Element model [13]. Practically our method has been implemented in C++ using Vega FEM library [13]. The following parameters are used: $k = 20$, time step $= 0.1$, the simulation ends when the mean distance between the current and the final point is less than 0.1 mm (note that the size of the pixel in our ultrasound images is 0.18mm). An exemple of convergence of the method is shown in Fig.2.

5   Results

Experiments have been conducted on several ultrasound sequences of natural speech. The size of the images is 532x434 while the size of the pixel is 0.18mm.

Method Accuracy Relative to the Point Location
Leave-one-out cross validation is used to evaluate the accuracy of the method. We took the case where a minimal distance $d = 15$ pixels is applied. We have alternatively removed every point $\mathbf{P}_F^n$ of the mechanical model and computed the euclidean distance between this point moving without constraint and $\mathbf{P}_F^n$. Fig. 2 shows an illustration of this method: light blue squares represent the position of the point calculated by the mechanical model, while pink squares represent the final position of the point, estimated with Farneback method. Fig. 3.a shows the error for every point. Two points display an error superior to 3 mm (#1 and #15). They correspond to locations close to dark areas where the computed variance is high. The other distances are low and show that our model is stable relatively to the constraint point locations. Fig. 3.a also shows that the median error is 0.7 mm: this range of accuracy is quite compatible with the dimensions of our target, namely building a dynamic model of the vocal tract from US images.

Method Accuracy Relative to the Number of Points
Our method also depends on the number N of points tracked by Farneback's method. We have tested the influence of this number N by applying our method with two different minimal distances $d = 30$ pixels and $d = 15$ pixels. Fig. 3.b shows two tongue contours computed with both datasets. There is a slight difference due to the too low constraint number with $d = 30$, which is not enough to guide the model. More accurate results are obtained when N increases ($d = 15$). However, we cannot use too small d distances since it may produce mechanical incoherences.

Method Validation on Ultrasound Sequences
Fig. 4 shows nine images extracted every over image in one ultrasound sequence. Upper tongue contours have been computed with our method. They have been

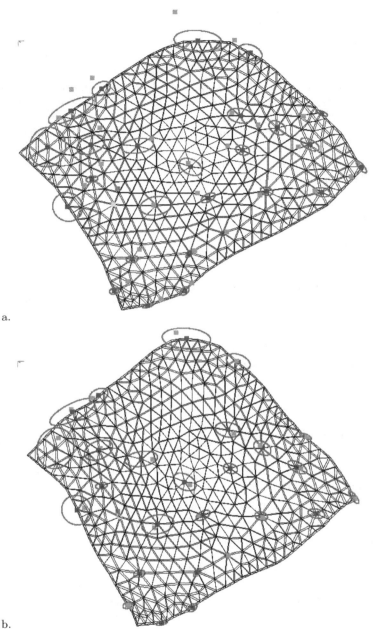

a.

b.

Fig. 2. Convergence of the biomechanical model: (a) initial position: points detected in the first image are in blue and their associated convergence is in red. They are integrated into the mesh. Green points are the points detected in the final image (b): Position of the mesh after convergence. All the green points are now within the covariance areas of the blue points.

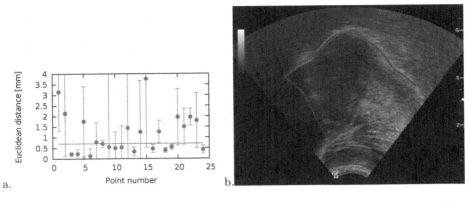

Fig. 3. Euclidean error for every tracked points. Median (blue line) is 0.71 mm. The largest eigenvalue of Λ_F^n is also displayed in green with a 50% scale. (b)Influence of d on the accuracy: more accurate results are obtained with d=15 pixels (red curve) than with d=30 pixel (red).

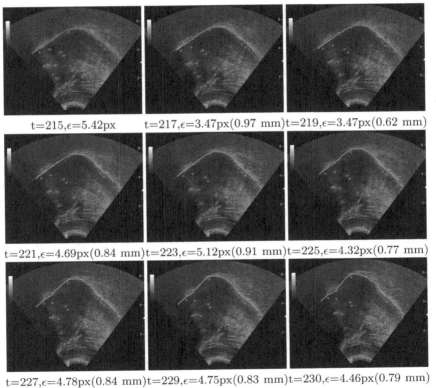

Fig. 4. Ultrasound image sequence: initial contour is in red, tracked points are with red circles and computed contours are in green. The errors ϵ show the mean distance (expressed in pixels and millimiters) between our method and manually segmented contours.

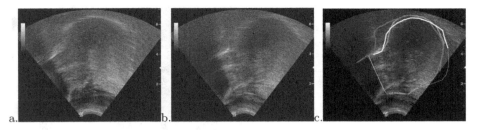

Fig. 5. a,b: Snaphots 1 and 15 of a 15 frames tongue sequence with a rearword motion. c: Tongue tracking with different methods: our method (white), Tang's method (red), CPD method (green). The points tracked in the sequence are drawn with red circles.

compared with a ground truth, which is a sequence of manually segmented contours. The errors ϵ between both contours for each image have been computed and are displayed in pixels and converted into millimeters. The errors remain below 1mm.

Comparison with Existing Methods

In order to prove the effectiveness of our approach, we have compared our method with the state-of-the art method for tongue tracking [15]. We also compare it with the well established CPD method which enables point set non-rigid registration (see [9] and the publicly available code). These methods were compared to manually detected contours on a challenging sequence with moving upward compression motion of the tongue (Fig.5, top). Comparison of these three methods are shown in (Fig.5, bottom). Tang's method (red curve) in unable to detect the right contour through the 15 frames sequence when sudden abrupt large motions occur. This is probably due to the temporal regularization term which is unable to cope with non-uniform temporal variations. CPD method (green curve) gives better results, but these results are dependent on several parameters (regularization and size of the Gaussian kernel). Our method (white curve) displays the best results despite the fact that there are few points on the right part of the tongue. This is due to the use of the elastic mechanical model which allows for better interpolation in areas where features are missing or sparse.

6   Conclusion

We have proposed a new approach for tongue tracking based on a biomechanical method. Using both contour points and inner points along with their uncertainties allows us to efficiently guide the mechanical model with a small number of points. Experiments show that the method is reliable and more efficient than existing methods to handle non uniform tongue movements. Future work will concern the pre-processing step in order to allow detection in few-textured areas and the incorporation of M-estimators to cope with possible outliers. We will also improve our mechanical model by adding collision detection with the palate in order to reproduce its crucial interaction with the tongue.

References

1. Aron, M., Roussos, A., Berger, M.-O., Kerrien, E., Maragos, P.: Multimodality Acquisition of Articulatory Data and Processing. In: 16th European Signal Processing Conference, EUSIPCO 2008 (2008)
2. Bressmann, T., Thind, P., Uy, C., Bollig, C., Gilbert, R.W., Irish, J.C.: Quantitative three-dimensional ultrasound analysis of tongue protrusion, grooving and symmetry: data from 12 normal speakers and a partial glossectomee. Clin. Linguist. Phon. 19(6-7), 573–588 (2005)
3. Buchaillard, S., Perrier, P., Payan, Y.: A biomechanical model of cardinal vowel production: muscle activations and the impact of gravity on tongue positioning. Journal of the Acoustical Society of America 126(4), 2033–2051 (2009)
4. Farhadloo, M.: Miguel Á.: Carreira-Perpiñán. Learning and adaptation of a tongue shape model with missing data. In: ICASSP, pp. 3981–3984 (2012)
5. Farnebäck, G.: Two-frame motion estimation based on polynomial expansion. In: Bigun, J., Gustavsson, T. (eds.) SCIA 2003. LNCS, vol. 2749, pp. 363–370. Springer, Heidelberg (2003)
6. Gerard, J.M., Ohayon, J., Luboz, V., Perrier, P., Payan, Y.: Non-linear elastic properties of the lingual and facial tissues assessed by indentation technique. Application to the biomechanics of speech production. Medical Engineering and Physics 27(10), 884–892 (2005)
7. Geuzaine, C., Remacle, J.-F.: Gmsh: A 3-D finite element mesh generator with built-in pre- and post-processing facilities. International Journal for Numerical Methods in Engineering 79, 1309–1331 (2009)
8. Haouchine, N., Dequidt, J., Peterlik, I., Kerrien, E., Berger, M.-O., Cotin, S.: Image-guided Simulation of Heterogeneous Tissue Deformation For Augmented Reality during Hepatic Surgery. In: ISMAR - IEEE International Symposium on Mixed and Augmented Reality 2013, Adelaide, Australia (October 2013)
9. Jian, B., Vemuri, B.C.: Robust point set registration using gaussian mixture models. IEEE Trans. Pattern Anal. Mach. Intell. 33(8), 1633–1645 (2011)
10. Li, M., Kambhamettu, C., Stone, M.: A level set approach for shape recovery of open contours. In: Narayanan, P.J., Nayar, S.K., Shum, H.-Y. (eds.) ACCV 2006. LNCS, vol. 3851, pp. 601–611. Springer, Heidelberg (2006)
11. Noble, J.A., Boukerroui, D.: Ultrasound image segmentation: a survey. IEEE Transactions on Medical Imaging 25(8), 987–1010 (2006)
12. Roussos, A., Katsamanis, A., Maragos, P.: Tongue tracking in ultrasound images with active appearance models. In: ICIP, pp. 1733–1736 (2009)
13. Sin, F., Schroeder, D., Barbic, J.: Vega: Non-linear fem deformable object simulator. Comput. Graph. Forum 32(1), 36–48 (2013)
14. Stone, M., Stock, G., Bunin, K., Kumar, K., Epstein, M., Kambhamettu, C., Li, M., Prince, J.: Comparison of speech production in upright and supine position. Journal of The Acoustical Society of America 122(1) (2007)
15. Tang, L., Bressmann, T., Hamarneh, G.: Tongue contour tracking in dynamic ultrasound via higher-order MRFs and efficient fusion moves. Medical Image Analysis 16(8), 1503–1520 (2012)
16. Wuhrer, S., Lang, J., Shu, C.: Tracking complete deformable objects with finite elements. In: 3DIMPVT, pp. 1–8 (2012)

Intra-Operative Registration
for Stereotactic Procedures Driven by a
Combined Biomechanical Brain and CSF Model

Alexandre Bilger[1], Éric Bardinet[2], Sara Fernandez-Vidal[2], Christian Duriez[1],
Pierre Jannin[3], and Stéphane Cotin[1]

[1] Inria Lille, Nord Europe Research Centre
[2] Centre de Recherche de l'Institut du Cerveau et de la Moelle épiniere,
UMR-S975, Paris, Inserm
[3] Equipe Medicis, U1099 LTSI, Université Rennes I

Abstract. During stereotactic neurosurgery, the brain shift could affect
the accuracy of the procedure. However, this deformation of the brain is
not often considered in the pre-operative planning step or intra-operatively,
and may lead to surgical complications, side effects or ineffectiveness. In
this paper, we present a method to update the pre-operative planning
based on a physical simulation of the brain shift. Because the simulation
requires unknown input parameters, the method relies on a parameter estimation process to compute the intracranial state that matches the partial data taken from intra-operative modalities. The simulation is based on
a biomechanical model of the brain and the cerebro-spinal fluid. In this paper, we show on an anatomical atlas that the method is numerically sound.

1 Introduction

Stereotactic neurosurgery requires a high precision for target location and target
definition. For example, one of the main targets in deep brain stimulation, the
subthalamic nucleus, is about 9 millimeters long. However, a combination of
deformation and motion of the brain, known as brain shift, can occur during the
procedure, depending on the surgical technique and the patient's brain anatomy
and pathology. In stereotactic procedures, it may cause a displacement of the
target or other structures. Elias et al.[4] reported an anterior commissure shift
up to 5.67 mm with a mean of 0.98 mm, and higher shift values for the frontal
pole.

Because brain shift is a major source of errors when applying the planned
strategy intra-operatively, several groups have studied the problem of brain shift
compensation using a deformable model, instead of using a fully intensity-based
method. Skrinjar et al. [8] presented a method to deform pre-operative data according to a partial intra-operative brain surface, captured by a stereo camera,
while Chen et al. [3] and Audette et al.[1] used a laser range scanner. However
the equipment is not standard and requires a large working space without occlu-
sion, which is not always the case in neurosurgery. The work of Wittek et al. [10]

F. Bello and S. Cotin (Eds.): ISBMS 2014, LNCS 8789, pp. 76–85, 2014.
© Springer International Publishing Switzerland 2014

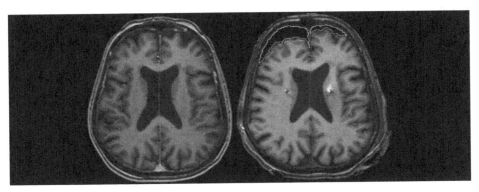

Fig. 1. Pre-operative (left) MR image without brain shift, and post-operative (right) MRI showing asymmetric brain shift. The green and pink contours is a segmentation of the pneumocephalus (air inside skull). The pink line represents the surface of the brain in contact with air.

is very similar to [8] but they used intra-operative MRI to guide the deformable model and took extra care on the complex model of the brain tissue deformation. However [8] and [10] do not model physically the brain shift phenomenon, because they include artificial forces by adding virtual springs or constraints between pre- and intra-operative control points. None of these methods accounts for the effects of gravity, interaction of CSF and brain tissue and loss of CSF. On the contrary, Chen et al. [3] used a computational model which accounts for CSF loss and gravity. They pre-operatively built a statistical atlas of deformation to solve the inverse problem intra-operatively.

We propose a new method to take into account brain shift during a stereotactic neurosurgical procedure. It relies on high resolution pre-operative MR image and intra-operative image (CT scan) of the patient's brain, acquired once the brain has shifted, but the method could be extended to other modalities such as intra-operative MRI. The pre-operative data are deformed according to a physical model of brain shift, mainly based on a cerebro-spinal fluid (CSF) interaction with the brain tissue, gravity and brain-skull interactions. To best fit the intra-operative configuration, the input parameters of the brain shift model are estimated intra-operatively with a measure of similarity with a series of pre-computed deformed configurations. The output of this inverse problem gives parameters used to compute the intra-operative configuration, as detailed as the pre-operative data. With this information, the surgeon is able to update the pre-operative planning data (target coordinates, trajectory angles etc).

The following section describes the brain shift simulation and how we use it to solve the intra-operative registration based on a physical parameters estimation. Section 3 presents the results and the numerical validation on an anatomical atlas. Section 4 concludes and addresses future steps for the method.

2   Materials and Methods

The overall pipeline of our method is detailed in the figure 2. This section presents the model of brain shift used in the simulations and the parameters estimation (the highlighted steps in the figure 2). The other aspects are not treated in this paper as we focus on the numerical validation.

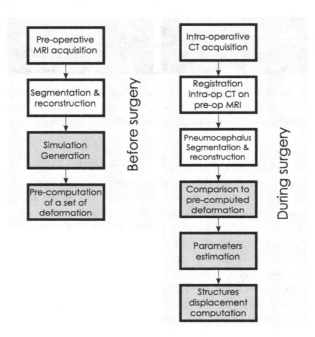

Fig. 2. Steps of our method (read from left to right, then from top to bottom): the acquired pre-operative MRI enables to generate a simulation. A set of deformation is pre-computed. During the surgery, the pneumocephalus is extracted from the intra-operative CT, rigidly registered to the pre-operative data to be in the same space and compared to the set of pre-computed deformation. Finally, the displacement of the structures of interest is deduced. Among these steps, the most time consuming processes are done before the surgery.

2.1   Presentation of the Model

The brain shift simulation relies on a physics-based model of the brain tissue deformation, the contact response with the skull and the falx cerebri, and the interaction with the CSF. It is based on the work of [2].

The brain deformation is computed using a non-linear geometric finite element method, with a linear constitutive law[5]. This allows for rotation in the model, while relying on a linear expression of the stress-strain relationship. The viscous part of the brain behavior is omitted, because we focus on the static equilibrium after the brain has shifted. The equation relating the external forces f to the nodal displacements u can be written as $f = K(u)u$, with K the stiffness matrix depending on u. Both hemispheres are meshed independently with linear tetrahedrons.

When the brain deforms, the simulation algorithm solves the contacts with the skull and the falx cerebri, considered as rigid. Unilateral interpenetration constraints are created when a collision is detected. The area near the brainsterm is assigned fixed Dirichlet boundary conditions to impose a null displacement constraint. The contacts are solved with Lagrangian multipliers. Figure 3 shows the boundary conditions of the model.

The main cause of brain shift is a CSF loss [9]. The resulting force of the surrounding fluid acting on the brain tissue acts against the weight, so that the brain is balanced. A CSF loss causes a fluid forces decrease, breaking the balance and leading to a brain shift. In order to cause a brain shift in our simulation, we include a model of CSF forces:

$$\mathbf{f}_{CSF} = \iint_S \rho \, g \, h(P) \, \mathrm{d}\mathbf{S}$$

where ρ is the density of CSF ($1000 \; kg/m^3$), g is the norm of the gravity and h is the distance between a point P on the mesh and the fluid surface. The force $\mathbf{f}_{CSF}$ is applied onto every immersed triangle S of the brain surface mesh. Asymmetric brain shift is observed on post-operative images (see Fig. 1). To handle this property, two independent models of fluid forces are present in the simulation, each acting independently on each hemispheres. With this model, the deformation depends also on the patient's head orientation compared to gravity direction, mechanical properties of the tissue and patient variability.

To recover the true rest configuration of the brain before applying gravity and external forces on the finite element model, we solve an inverse problem using an iterative geometric algorithm [7]. The algorithm is used with a static solver. The iterations are smoothed to handle the geometrical non-linearities of the deformation model.

2.2 Non-rigid Registration Based on Physical Parameters Estimation

Parameters to Estimate. As explained in section 2.1, the brain shift model depends on the following parameters: patient's head orientation compared to gravity direction, patient variability (geometry), mechanical properties of the tissue and the amount of CSF lost during the procedure. The two first parameters are known: we can measure the orientation of the patient's head and the geometry is acquired by the segmentation of patient images. The mechanical parameters of the brain tissue have been estimated by several groups using different techniques, but there is no consensus and there is no direct measurement

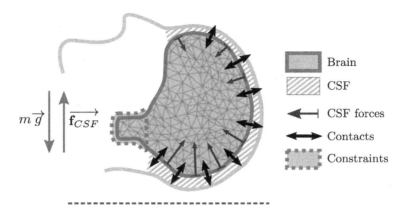

Fig. 3. Schematic representation of the brain model and the set of constraints describing its interactions with the environment. In this pre-operative configuration, the patient is in the supine position (see the gravity vector direction $\vec{g}$). The cerebro-spinal fluid surrounds the brain tissue and acts on it with pressure forces. The resultant force balances the weight (see the $\overrightarrow{\mathbf{f}_{CSF}}$ compared to the weight force $m\vec{g}$). The brain is under a null displacement constraint near the brainstem area. The illustration also shows the contacts between the brain and the skull.

technique personalized for a patient. Although it would be interesting to add the mechanical properties to the parameters to estimate, the deformation is also controlled by the CSF volume lost, meaning that these two parameters cannot be dissociated. That is why the mechanical properties are fixed to the value used in [6]. Moreover, the mechanical parameters are not the physical cause of the brain shift phenomenon. We will show that an acceptable error in the mechanical parameters does affect the parameter estimation, but not the output of our algorithm. Finally, only the CSF volume lost is estimated. As we model both hemispheres independently, two quantities have to be estimated.

Error Minimization. In our algorithm, the only unknown input parameters are two volumes of CSF lost during the surgery. To estimate these parameters, we introduce a measure of similarity, to compare the final deformed geometry of the brain, obtained at equilibrium after applying a CSF loss in a simulation, and the visible surface of the brain, extracted from an intra-operative modality. The similarity between these two objects is defined with a least-squares approach:

$$d(S, M) = \sum_{x_S \in S} \|x_S - p(x_S, M)\|^2 \tag{1}$$

with x_S a point on the surface S of the intra-operative brain, $p(x_S, M)$ is the projection of the point x_S on the surface M of the simulated brain. The measure is normalized by dividing by the number of points in S, and represents the average distance between both surfaces. The figure 4 shows a representation of such a projection in image space.

Our model is consistent with the correlation between the CSF loss and the amount of deformation, studied by [4]. The displacement norm of a point in the brain is an increasing function of the CSF volume lost. This property guarantees the convexity of the error as a function of the two CSF volume losses (see an example in Fig. 5): a local minimum is a global minimum.

Even an efficient optimization method for our problem would require at least several tens of iterations to converge to an estimation of the parameters. Each iteration requires to compute a direct simulation, which takes time (more than 5 minutes) and therefore is not compatible with an application during a neurosurgical procedure. To speed up the intra-operative computation, we pre-compute before the surgery a set of brain configurations, deformed by a simulated brain shift for a regular sampling of the parameters domain. Then, during the procedure, a mesh is extracted from an intra-operative image, and compared to all the pre-computed deformed configurations. The configuration with the smallest error corresponds to the registered brain.

Finally, the displacement field enables to compute rapidly the displacement and deformation of other structures such as the ventricles, the target or the blood vessels.

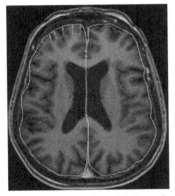

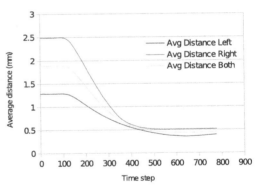

(a) Measure of similarity: the intra-operative triangular surface (pink) is projected on the simulated brain surface (yellow). The measure is the average distance of projection.

(b) At each time step of a simulation, the average distance between the simulated brain and the intra-operative data is measured. In this example, the simulation is parameterized with the same values used to generate the intra-operative data, that is why the distance converges to a small value.

Fig. 4. Mesure of similarity between the simulated brain state and intra-operative data

3 Results

To validate our method, we first apply it on a template containing high-fidelity models of the brain, skull, skin, ventricles and blood vessels. In our tests, both hemispheres are meshed with approximately 10,000 tetrahedrons. The Young's modulus E is set to 3000 Pa and Poisson's ratio to 0.45 according to [6].

In this section, we show that our method is numerically sound. A virtual brain shift, obtained with fixed chosen parameters, is applied on the template. The obtained configuration will be used as a synthetic of 3D intra-operative data. These data are then used in the parameters estimation process to prove the consistency of the method: the estimated parameters are compared to the parameters used to generate the data.

Finally, we also show that if the Young's modulus differs slightly from the mechanical parameters used to generate the synthetic intra-operative mesh, our algorithm still register the intra-operative brain with a minimal error.

Data Generation. With given input parameters V_L and V_R (CSF loss volume for both sides), a deformation is computed with the brain shift simulation: the pre-operative configuration undergoes a CSF leak leading to a brain shift. After the brain shift, when the deformed brain is at equilibrium, the surface in contact with air is extracted, i.e. the same data we extract from intra-operative patient data (see pink line in Fig. 1). The surface is uniformly remeshed with a marching cube algorithm to mimic data coming from 3D images, and noise is added, to meet as much as possible the intra-operative conditions. Finally, the parameters leading to this surface are estimated.

Validation Protocol. Multiple intra-operative data were generated in order to validate different scenarios: symmetric brain shift(#1), asymmetric brain shift (#2), unknown mechanical parameters for both symmetric (#3) and asymmetric brain shifts (#4). The parameters (CSF loss volumes and Young's modulus) of each scenario are summarized in the table 1.

Table 1. Different scenarios of brain shift used to validate the method. The scenarios #2 and #4 mimic an asymmetric brain shift. #3 and #4 use data generated with mechanical parameters different from the minimization process.

	Input parameters				Optimization output			
	Volumes(cm^3)		Young's modulus(Pa)		Volumes(cm^3)		Average distance (mm)	
Scenario	Left	Right	E_{data}	E_{simu}	Left	Right	Left	Right
#1	30	30	3000	3000	30	30	0.4	0.4
#2	10	20	3000	3000	10	20	0.5	0.5
#3	30	30	3000	4000	40	40	0.4	0.4
#4	10	20	3000	4000	15	25	0.6	0.5

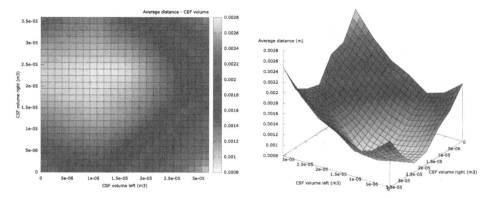

Fig. 5. The CSF loss volume are varying (X and Y axis). The colors represents the average distance between the equilibrium state of the deformed brain caused by a CSF loss (x,y). The minimum is located at $(1 \times 10^{-5}, 2 \times 10^{-5})$, which is the set of parameters used to generate the intra-operative data (scenario #2).

Error Measurement. For each scenario, a set of deformation is computed, and compared to the synthetic intra-operative model, in order to find the minimum distance. The figure 5 shows the average distance between data generated with the scenario #2 and the simulated brain shifts, with CSF losses varying between 0 and 30cm^3.

The table 1 summarizes the error between the values used to generate the synthetic intra-operative model, and the values estimated by the minimization process. These values are comparable only if the mechanical parameters used for generation and optimization process. To measure errors if the mechanical parameters are different, we compare directly the brain geometry with the measure of similarity presented in the section 2.2.

The data in the table 1 show that, in the scenarios #1 and #2, the volume parameters are the same in input and output (the error in the measure of similarity comes from the remeshing and noise in the intra-operative data). In scenarios #3 and #4, where the brain is stiffer in the parameters estimation, the volumes estimated are greater, as expected, but the brain deformed with these parameters still matches the intra-operative data. We conclude that the Young's modulus does not have an effect on the registration output.

The figure 6 shows an example of the difference between the pre-operative configuration of the brain and the deformed brain fitting the intra-operative data. With our method, we are able to estimate the CSF volume lost, which gives a registration of the brain if used in the brain shift model. The entire process is based on a physical model, and no artificial forces are needed. The registration algorithm also provides the deformation and displacement of brain structures thanks to the displacement field.

Fig. 6. 3D view of the mechanical model of the brain (in blue, only cut for visualization purpose). In pink, the mesh corresponding to intra-operative data. Left: the mechanical model is in its pre-operative configuration. Right: The brain fits the intra-operative data after a deformation.

4   Conclusion

In this paper, we presented a method to register the pre-operative configuration of the brain onto the intra-operative state. The registration is entirely based on physics and handle patient-specific geometry, patient's head orientation and asymmetry in the deformation. We showed the numerical consistency of the method on 4 scenarios involving symmetry/asymmetry and mechanical parameters changes. The algorithm provides also the physical displacement and deformation of other structures such as the blood vessels or the target. The method is independent from the constitutive law of the brain, and could be improved using a more accurate law. The next step of this work is to validate it on a series of patient.

Acknowledgment. The authors would like to thank the French Research Agency (ANR) for funding this study through the ACouStiC project.

References

1. Audette, M.A., Siddiqi, K., Ferrie, F.P., Peters, T.M.: An integrated range-sensing, segmentation and registration framework for the characterization of intra-surgical brain deformations in image-guided surgery. Computer Vision and Image Understanding 89(2-3), 226–251 (2003)
2. Bilger, A., Duriez, C., Cotin, S.: Computation and visualization of risk assessment in deep brain stimulation planning. Studies in Health Technology and Informatics 196, 29–35 (2014)
3. Chen, I., Coffey, A.M., Ding, S., Dumpuri, P., Dawant, B.M., Thompson, R.C., Miga, M.I.: Intraoperative brain shift compensation: accounting for dural septa. IEEE Transactions on Bio-medical Engineering 58(3), 499–508 (2011)

4. Elias, W.J., Fu, K.M., Frysinger, R.C.: Cortical and subcortical brain shift during stereotactic procedures. Journal of Neurosurgery 107(5), 983–988 (2007)
5. Felippa, C., Haugen, B.: A unified formulation of small-strain corotational finite elements: I. Theory. Computer Methods in Applied Mechanics and Engineering 194(21-24), 2285–2335 (2005)
6. Ferrant, M., Nabavi, A., Macq, B., Black, P.M., Jolesz, F.A., Kikinis, R., Warfield, S.K.: Serial registration of intraoperative MR images of the brain. Medical Image Analysis 6(4), 337–359 (2002)
7. Sellier, M.: An iterative method for the inverse elasto-static problem. Journal of Fluids and Structures 27(8), 1461–1470 (2011)
8. Skrinjar, O., Nabavi, A., Duncan, J.: Model-driven brain shift compensation. Medical Image Analysis 6(4), 361–373 (2002)
9. Slotty, P.J., Kamp, M.A., Wille, C., Kinfe, T.M., Steiger, H.J., Vesper, J.: The impact of brain shift in deep brain stimulation surgery: observation and obviation. Acta Neurochirurgica (August 2012)
10. Wittek, A., Miller, K., Kikinis, R., Warfield, S.K.: Patient-specific model of brain deformation: application to medical image registration. Journal of Biomechanics 40(4), 919–929 (2007)

3D CFD in Complex Vascular Systems: A Case Study⋆

Olivia Miraucourt[1,2], Olivier Génevaux[3], Marcela Szopos[4], Marc Thiriet[5],
Hugues Talbot[2], Stéphanie Salmon[1], and Nicolas Passat[6]

[1] Université de Reims Champagne-Ardenne, LMR, France
[2] Université Paris-Est, ESIEE, CNRS, LIGM, France
[3] Université de Strasbourg, CNRS, ICube, France
[4] Université de Strasbourg, CNRS, IRMA, France
[5] Université Paris 6, CNRS, LJLL, France
[6] Université de Reims Champagne-Ardenne, CReSTIC, France

Abstract. Modeling the flowing blood in vascular structures is crucial to perform *in silico* simulations in various clinical contexts. This remains however an emerging and challenging research field, that raises several open issues. In particular, a compromise is generally made between the completeness of the simulation and the complicated architecture of the vasculature: reduced order simulations (lumped parameter models) represent vascular networks, whereas detailed models are devoted to small regions of interest. However, technical improvements enable targeting of compartments of the blood circulation rather than focusing on vascular branched segments. This article aims at investigating the cerebral flow in the entire venous drainage that can be reconstructed from medical imaging.

1 Introduction

The cerebral vasculature is a highly complex three-dimensional network, composed by three successive compartments: arteries, capillaries and veins. The arterial cerebral vasculature is characterized by a relatively invariant anatomic pattern [1, 2]. The major part of the current literature on macro/mesoscopic modeling and simulation of blood flow in the brain has focused on this first compartment [3]. By contrast, venous cerebral system was rarely studied and still remains not well understood. In particular, several difficulties are to be considered: the lack of parallelism between arterial and venous circulations; the asymmetric and considerably more various pattern of the venous network compared to the arterial; the highly individual variations of the venous outflow.

In this challenging context, flowing blood simulation is highly relevant for various clinical purposes. It can allow diagnosis and follow-up of aneurysms [4], or predict the effects of stenting [5] or coiling [6] procedures. In order to be actually useful, blood flow modeling has to fit at best the physical rules of fluid dynamics, with respect to

⋆ This research was funded by a grant from the *Région Champagne-Ardenne* and by *Agence Nationale de la Recherche* (Grant Agreement ANR-12-MONU-0010). MRI images were provided by the In Vivo Imaging Platform of Université de Strasbourg. Computing resources were provided by the HPC Center of Université de Reims Champagne-Ardenne.

physiological assumptions. As a consequence, by opposition to many clinical applications where virtual reality provides satisfactory paradigms, it is mandatory to consider mathematical approaches relying on numerical analysis [7]. This strategy presents a higher computational cost, but also a higher – and crucial – physical reliability.

Computational fluid dynamics has evolved with the development of new numerical schemes and high performance computing. However, despite many breakthroughs, open problems still exist. They derive from computational cost issues, but also from the difficulty to accurately model the multiple physiological assumptions. In this methodological framework, two dual strategies are generally considered.

The first strategy consists of performing *reduced order* simulations, i.e., to consider 1D models of the 3D vessels. The strong decrease of space complexity allows the numerical solvers to deal with complex vascular networks, and to provide accurate – yet macroscopic – behavior of the global flowing blood. This paradigm has been considered to handle the vascular networks of specific organs [8], but also of the whole human body [9–11]. However, local values of vascular resistances and compliances remain, in general, unknown in the context of personalized studies.

The second strategy consists of performing *complete* computational fluid dynamics simulations, i.e., to consider the 3D geometry of the vessels, and to simulate fine blood flow patterns. In particular, it has been demonstrated [12] that a satisfactory agreement is obtained by 1D models during diastole, but significant differences appear during systole, partially explained by the inability of the currently used 1D models to take into account secondary flow features, vessel curvature, etc. In 3D models, the huge size of the induced equation systems generally leads to focus on vessel samples. Classical use cases are then related to large vessels [13] or to vascular pathologies [4, 14] where the considered vascular geometries are vessel segments. Unfortunately, sampling the vessels induces a bias that alters the correctness of the flow information [15].

This motivates the development of simulation strategies that accurately handle both complicated vascular geometries and 3D blood flow models. Approaches based on mixed 3D/1D computational fluid dynamics models [16] are a first – yet partial – step towards this goal. (The use of 3D/1D coupled models allows to study how local and global hemodynamics phenomena are linked, for instance, the carotids can be interfaced with the more complex arterial network [17, 18].) Regarding a fully 3D context, few attempts have been proposed for the cerebral arterial [19] and venous [20] networks. Both were preliminary, and the latter essentially dealt with the initialization of the simulation, i.e. a stationary Stokes problem.

Our efforts toward this goal have been carried out by following three fundamental guidelines: (i) developing/using open-source softwares, in order to have a full handling of the underlying methods, and to provide guarantees in terms of current and future availability as well as reproducible research; (ii) proving the numerical correctness of the developed solvers, giving quality guarantees both in terms of reliability, reproducibility and incrementality, in contrast to many previous approaches based on commercial software packages; and (iii) considering a cautious physically-oriented modeling of the flowing blood behavior, to obtain not only realistic, but also reliable – and thus hopefully useful – results.

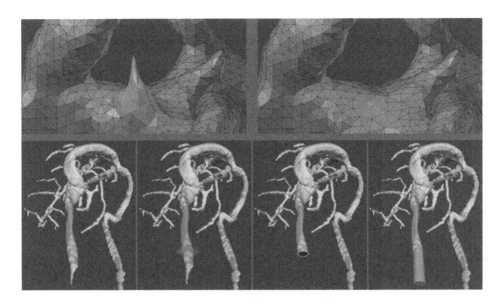

Fig. 1. First row: geometry correction of a 2D mesh. Second row: processing of the inputs and outputs (identification, cutting and extrusion).

The article is structured as follows. Sec. 2 summarizes the chain to obtain a realistic geometry. Sec. 3 is devoted to the modeling of the venous network in order to obtain the adequate equations to solve. In Sec. 4, we give details on the considered numerical approaches and algorithms. We present in Sec. 5 numerical results on analytical solutions and on a realistic case. We gather in Sec. 6 some concluding remarks.

2   Geometric Modeling

The first step of our scheme deals with image processing and mesh generation issues. It takes as input a 3D MRA (Magnetic Resonance Angiography) data, whose resolution is sufficient at a macroscopic scale. Images were acquired on a whole-body scanner (Siemens Magnetom Verio $3.0\,T$, gradient slope $= 200\,T/m/s$, flow encoding sequence, TR $= 42.7\,ms$, TE $= 6.57\,ms$, resolution $= 0.4 \times 0.4 \times 0.8\,mm^3$).

An example-based segmentation method [21] is used to extract the venous 3D digital volume. Marching-cubes is then used to compute its boundary as a 2D meshed surface. These meshes cannot be used straight-away to generate the 3D computational fluid dynamics meshes. Indeed, (i) local geometric corrections are needed before being suitable for simulation, and (ii) it is required to identify domain input/outputs. The first task is crucial, as geometric artifacts can remain locally despite the efforts put into the segmentation, such as spikes off the main surface due to small branching vessels, undesired small branches, small disconnected components, etc. The second task is required so one can latter setup appropriate boundary conditions for the numerical model on each venous tree tip (Fig. 1). Due to the necessity to interactively inspect the mesh, we tackled these tasks with a workflow supported by a combination of off-the-shelf tools such as the BLENDER modeling software and appropriately developed software. The former is geared towards general mesh editing to fix areas that are deemed to be problematic while the latter is more dedicated to perform problem-specific tasks efficiently, for instance inputs and outputs labeling and reshaping. Finally, we smooth the 2D mesh with MMGS before building the 3D tetrahedral simulation mesh with MMG3D[1]. The size of mesh, with respect to the number of degrees of freedom, is chosen at this step.

3   Physical Modeling

Our aim is now to derive an appropriate model for the large cerebral veins located in the leptomeninges down to the extracranial drainage. At the macroscopic scale, the brain venous network (see illustration) is composed by – input – veins (vein of Galen (7); internal cerebral vein (8); basilar vein (9); superior cerebral veins (10); superior anastomotic veins (11)) draining the blood into the superior sagittal sinus (2) and the straight sinus (3), until their confluence (4). The blood then passes into the transverse (5) and sigmoid parts (6) of the lateral sinuses, and reaches an extracranial area, composed of the – output – internal jugular veins (1) [1].

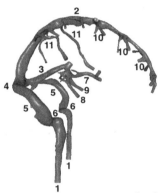

Concerning the flow in large/medium-sized cerebral veins, we adopt standard assumptions: (i) the blood density is constant; (ii) the flow is incompressible and isothermal. Blood is also supposed to be Newtonian [3, 22] as, in the absence of blood stagnation regions, rouleaux of red blood capsules are formed neither in the upstream microcirculation (particle train flow) nor in veins. Moreover, in such veins, flowing cells do not deform. Hence thixotropic, viscoelastic, and shear thinning features are omitted in the context of a drainage circuit that receives a steady, laminar flow at its entrance.

Another important issue is the relevance of using either a complex fluid-structure interaction model or a fluid model. Intracranial veins that run through the leptomeninges (mostly water) are quite constrained between a nearly incompressible brain and the rigid skull. They are thus not susceptible to experience large deformations, and are considered as rigid. (More generally neglecting wall motion has been shown to be an acceptable hypothesis when studying intracranial arterial blood flow [23].)

The boundary conditions are: (i) inflow: a steady profile (constant velocity of small magnitude, due to microcirculation exit, far from the thorax hence far from the effects of breathing and cardiac pumping [24]); (ii) outflow: homogeneous natural conditions; (iii) lateral boundary: no-slip condition, since walls are assumed to be rigid.

Computed orders of the Reynolds ($90 \rightarrow 232$), Stokes ($1.10 \rightarrow 3.84$) and Strouhal numbers ($0.013 \rightarrow 0.063$) show the potential importance of the convective forces and flow unsteadiness. In conclusion, the mathematical model that we consider is the Navier-Stokes equations with mixed boundary conditions

---

[1] MMGS and MMG3D are freely available at the following url:
http://www.ann.jussieu.fr/frey/software.html.

$$\begin{cases} \partial_t \mathbf{u} + \mathbf{u} \cdot \nabla \mathbf{u} - \nu \Delta \mathbf{u} + \nabla p = \mathbf{f} & \text{in } \Omega \times (0, T) \\ \nabla \cdot \mathbf{u} = 0 & \text{in } \Omega \times (0, T) \\ \mathbf{u} = \mathbf{g}_1 & \text{on } \Gamma_D \times (0, T) \\ \nu \dfrac{\partial \mathbf{u}}{\partial \mathbf{n}} - p\mathbf{n} = \mathbf{g}_2 & \text{on } \Gamma_N \times (0, T) \\ \mathbf{u}|_{t=0} = \mathbf{u}_0 & \text{in } \Omega \end{cases} \tag{1}$$

where $\Omega \subset \mathbb{R}^3$ is a bounded domain with a Lipschitz-continuous boundary Γ, $(0, T)$ is a finite time interval. We recall that $\mathbf{u}$ is the fluid velocity, p its pressure and ν its kinematic viscosity. We suppose that Γ consists of two measurable parts: Γ_D, where Dirichlet boundary conditions are imposed, and Γ_N, where Neumann boundary conditions are prescribed. We denote by $\mathbf{n}$ the unit outward normal vector to Ω on its boundary Γ. The Dirichlet boundary conditions correspond in the non-homogeneous case to a velocity $\mathbf{g}_1$, generally prescribed at the inlets, and in the homogeneous case to the non-slip boundary condition prescribed at the vessel wall; the function $\mathbf{g}_2$ corresponds to the normal component of the stress tensor, prescribed at the outlets.

4   Numerical Methods

To solve Eq. (1), we consider two numerical schemes presenting dual properties. The first [25, 3] considers a discretization involving a time-scheme based on the characteristics method and a spatial discretization of finite element type. Theoretically expected convergence in space for velocity and pressure error are of order 3 and 2 (resp. 2 and 1) for the L^2-norm (resp. the semi-norm H^1). The characteristics method is proven to be unconditionally stable and expected to be of order 1 in time.

The second method [26] belongs to the class of truly consistent splitting schemes [27], which decouples the computation of the velocity and pressure. It computes an approximate solution to the Navier-Stokes equations by solving at each time step a Poisson equation for the pressure and a heat equation for each component of the velocity. Its temporal accuracy was numerically assessed in 2D analytical cases [26] and observed to be of order 2 for both velocity and pressure. The spatial convergence was not evaluated, but second order accuracy for the space error is also expected.

Our implementations rely on the FREEFEM++ library [28], that enables to run algorithms on parallel architectures. All codes and details are freely available on-line[2].

We performed a validation on a gold-standard CFD problem with known analytical solutions [29]. The results (Fig. 2, log scale) confirm the numerical soundness of the schemes and our implementation. Our analysis shows that the second enjoys high-order convergence properties. However, the first remains faster in its current implementation.

5   Results

For our experiments on a real vascular model, a finite element mesh of 237 438 tetrahedra was generated. Since blood comes from microcirculation (quasi-steady/steady

---

[2] http://numtourcfd.univ-reims.fr

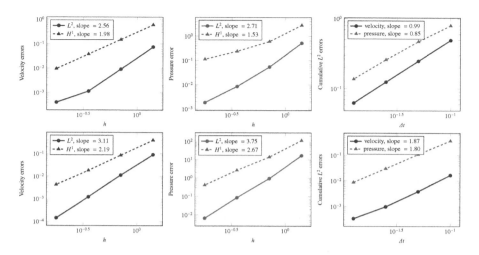

Fig. 2. Convergence curves of numerical errors. First row: [3]; second row: [26]. From left to right: space errors for velocity; space error for pressure; time errors for velocity and pressure.

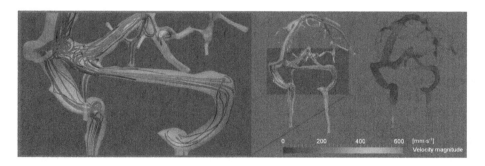

Fig. 3. Navier-Stokes simulation (iteration 250/1 000): streamlines and magnitude of velocity projected on the 2D surface

Stokes flow), we use a steady profile of small magnitude (28 mm/s), at the inflow. The no-slip condition was imposed on the wall boundaries, that are assumed to be rigid. For the outflow, we used a free traction boundary condition at the jugular vein extremities.

We performed 1 000 iterations, with a time-step equal to 10^{-3} s, in 14 hours on 16 processors, thus providing a simulation of 1 second of non-steady blood flow within cerebral venous network. Results obtained with this method are illustrated in Fig. 3–4.

Flows calculated at inlets and at outlets are identical (mass conservation property), with a value of 6 227 mm^3/s, corresponding to physiological data [2]. Complex pattern flows are present and an asymmetric behavior in the two outlet branches can be observed, due to the physiological asymmetric pattern of the venous network. The depicted pressure should be interpreted relative to a reference pressure, applied to the

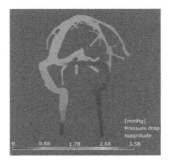

Fig. 4. Navier-Stokes simulation (iteration 250/1 000): magnitude of pressure drop field

models outlets; the obtained pressure drop of approximately 3.5 *mmHg* is in a physiological range [11]. It is worth noting that, since we use a rigid-walls model, the results are identical regardless of which value is used as a reference.

6   Conclusion

Many questions are raised by this work. Among the most crucial, one can cite the potential influence of the beating heart and the respiration on the outflow. These physiological phenomena should be taken into account in the outflow boundary conditions. In addition, the boundary conditions should take into account the position – horizontal or vertical – of the subject [30]. We also have to work on the collapsibility of the jugular veins. All these considerations may further lead to a better modeling.

Beyond the numerical validations with respect to analytical results, a main issue remains the validations in the real vascular network. If physical phantoms can be considered to reach that goal in the case of simple vascular structures, such approaches are hardly tractable for complex networks. The development of virtual angiography, that simulates the physical image acquisition sequences, in order to create realistic synthetic data, provides a promising perspective. The case of Computed Tomography Angiography has already been pioneered [31]. The case of MRA is an emerging research field that constitutes one of our very goals, and motivates our use of MRA images.

References

1. Schaller, B.: Physiology of cerebral venous blood flow: From experimental data in animals to normal function in human. Brain Res. Brain Res. Rev. 46, 243–260 (2004)
2. Stoquart-Elsankari, S., Lehmann, P., Villette, A., Czosnyka, M., Meyer, M.E., Deramond, H., Balédent, O.: A phase-contrast MRI study of physiologic cerebral venous flow. J. Cereb. Blood F. Met. 29, 1208–1215 (2009)
3. Formaggia, L., Quarteroni, A., Veneziani, A.: Cardiovascular Mathematics. MS & A, vol. 1. Springer (2009)
4. Cebral, J.R., Castro, M.A., Appanaboyina, S., Putman, C.M., Millan, D., Frangi, A.F.: Efficient pipeline for image-based patient-specific analysis of cerebral aneurysm hemodynamics: Technique and sensitivity. IEEE T. Med. Imaging 24, 457–467 (2005)

5. Larrabide, I., Kim, M., Augsburger, L., Villa-Uriol, M.C., Rüfenacht, D., Frangi, A.F.: Fast virtual deployment of self-expandable stents: Method and in vitro evaluation for intracranial aneurysmal stenting. Med. Image Anal. 16, 721–730 (2012)

6. Morales, H.G., Larrabide, I., Geers, A.J., Román, L.S., Blasco, J., Macho, J.M., Frangi, A.F.: A virtual coiling technique for image-based aneurysm models by dynamic path planning. IEEE T. Med. Imaging 32, 119–129 (2013)

7. Taylor, C.A., Figueroa, C.A.: Patient-specific modeling of cardiovascular mechanics. Annu. Rev. Biomed. Eng. 11, 109–134 (2009)

8. Ho, H., Mithraratne, K., Hunter, P.: Numerical simulation of blood flow in an anatomically-accurate cerebral venous tree. IEEE T. Med. Imaging 32, 85–91 (2013)

9. Reymond, P., Merenda, F., Perren, F., Rüfenacht, D., Stergiopulos, N.: Validation of a one-dimensional model of the systemic arterial tree. Am. J. Physiol. 297, 208–222 (2009)

10. Blanco, P.J., Leiva, J.S., Buscaglia, G.C.: A black-box decomposition approach for coupling heterogeneous components in hemodynamics simulations. Int. J. Num. Meth. Biomed. Eng. 29, 408–427 (2013)

11. Müller, L.O., Toro, E.F.: A global multiscale mathematical model for the human circulation with emphasis on the venous system. Int. J. Num. Meth. Biomed. Eng. (in press)

12. Xiao, N., Alastruey, J., Figueroa, C.A.: A systematic comparison between 1-D and 3-D hemodynamics in compliant arterial models. Int. J. Num. Meth. Biomed. Eng. 30, 204–231 (2014)

13. Camara, O., Mansi, T., Pop, M., Rhode, K., Sermesant, M., Young, A. (eds.): STACOM 2013. LNCS, vol. 8330. Springer, Heidelberg (2014)

14. Boissonnat, J.D., Chaine, R., Frey, P., Malandain, G., Salmon, S., Saltel, E., Thiriet, M.: From arteriographies to computational flow in saccular aneurisms: The INRIA experience. Med. Image Anal. 9, 133–143 (2005)

15. Sato, K., Imai, Y., Ishikawa, T., Matsuki, N., Yamaguchi, T.: The importance of parent artery geometry in intra-aneurysmal hemodynamics. Med. Eng. Phys. 30, 774–782 (2008)

16. Ho, H., Sorrell, K., Peng, L., Yang, Z., Holden, A., Hunter, P.: Hemodynamic analysis for transjugular intrahepatic portosystemic shunt (TIPS) in the liver based on a CT-image. IEEE T. Med. Imaging 32, 92–98 (2013)

17. Passerini, T., de Luca, M., Formaggia, L., Quarteroni, A., Veneziani, A.: A 3D/1D geometrical multiscale model of cerebral vasculature. J. Eng. Math. 64, 319–330 (2009)

18. Blanco, P.J., Pivello, M.R., Urquiza, S.A., Feijoo, R.A.: On the potentialities of 3D-1D coupled models in hemodynamics simulations. J. Biomech. 42, 919–930 (2009)

19. Mut, F., Wright, S., Ascoli, G., Cebral, J.R.: Characterization of the morphometry and hemodynamics of cerebral arterial trees in humans: A preliminary study. In: CMBE, pp. 87–90 (2011)

20. Miraucourt, M., Salmon, S., Szopos, M., Thiriet, M.: Blood flow simulations in the cerebral venous network. In: CMBE, pp. 187–190 (2013)

21. Dufour, A., Tankyevych, O., Naegel, B., Talbot, H., Ronse, C., Baruthio, J., Dokládal, P., Passat, N.: Filtering and segmentation of 3D angiographic data: Advances based on mathematical morphology. Med. Image Anal. 17, 147–164 (2013)

22. Thiriet, M.: Cell and Tissue Organization in the Circulatory and Ventilatory Systems. Springer (2011)

23. Sforza, D.M., Löhner, R., Putman, C., Cebral, J.R.: Hemodynamic analysis of intracranial aneurysms with moving parent arteries: Basilar tip aneurysms. Int. J. Num. Meth. Biomed. Eng. 26, 1219–1227 (2010)

24. Thiriet, M.: Biology and Mechanics of Blood Flows, part I: Biology of Blood Flows, part II: Mechanics and Medical Aspects of Blood Flows. Springer (2008)

25. Pironeau, O.: On the transport-diffusion algorithm and its applications to the Navier-Stokes equations. Numer. Math. 38, 309–332 (1982)

26. Sheng, Z., Thiriet, M., Hecht, F.: A high-order scheme for the incompressible Navier-Stokes equations with open boundary condition. Int. J. Numer. Meth. Fl. 73, 58–73 (2013)
27. Guermond, J.L., Shen, J.: A new class of truly consistent splitting schemes for incompressible flows. J. Comput. Phys. 192, 262–276 (2003)
28. Hecht, F.: New development in Freefem++. J. Num. Math. 20, 251–265 (2012)
29. Ethier, C.R., Steinman, D.A.: Exact fully 3D Navier-Stokes solutions for benchmarking. Int. J. Numer. Meth. Fl. 19, 369–375 (1994)
30. Gisolf, J., van Lieshout, J.J., van Heusden, K., Pott, F., Stok, W.J., Karemaker, J.M.: Human cerebral venous outflow pathway depends on posture and central venous pressure. J. Physiol. 560, 317–327 (2004)
31. Ford, M.D., Stuhne, G.R., Nikolov, H.N., Habets, D.F., Lownie, S.P., Holdsworth, D.W., Steinman, D.A.: Virtual angiography for visualization and validation of computational models of aneurysm hemodynamics. IEEE T. Med. Imaging 24, 1586–1592 (2005)

Computational Stent Placement in Transcatheter Aortic Valve Implantation

Christoph Russ[1], Raoul Hopf[2], Simon H. Sündermann[4], Silvia Born[3],
Sven Hirsch[1], Volkmar Falk[4], Gábor Székely[1], and Michael Gessat[1,3]

[1] Computer Vision Laboratory, ETH Zurich
Sternwartstrasse 7, 8092 Zurich, Switzerland
russc@ethz.ch
[2] Institute of Mechanical Systems, ETH Zurich
Tannenstrasse 3, 8092 Zurich, Switzerland
[3] Hybrid Laboratory for Cardiovascular Technologies, University of Zurich
Rämistrasse 101, 8091 Zurich, Switzerland
[4] Division of Cardiovascular Surgery, University Hospital
Rämistrasse 101, 8091 Zurich, Switzerland

Abstract. Transcatheter aortic valve implantation (TAVI) is a minimally invasive procedure to treat severe aortic stenosis in patients with a high risk for conventional surgery. In-silico experiments of stent deployment within patient-specific models of the aortic root have created an opportunity to predict stent behavior during the intervention. Current limitations in procedure planning are a primary motivator for these simulations. The virtual stent placement preceding the deployment phase of such experiments has major influence on the outcome of the simulation, but only received little attention in literature up to now. This work presents a methodical approach to patient-specific planning of placement of self-expanding stent models by analyzing experimental outcomes of different sets of boundary conditions constraining the stent. As a results, different paradigms for automated or expert guided stent placement are evaluated, which demonstrate the benefits of virtual stent deployment for intervention planning. To build a predictive planning pipeline for TAVI we use an automatic segmentation of the aorta, aortic root and left ventricle, which is converted to a finite element mesh. The virtual stent is then placed along a guide wire model and deployed at multiple locations around the aortic root. The simulation has been evaluated using pre- and post-interventional CT scans with an average relative circumferential error of 4.0% ($\pm$2.5%), which is less than half of the average difference in circumference between individual stent sizes (8.6%). Our methods are therefore enabling patient-specific planning and provide better guidance during the intervention.

Keywords: stent placement, virtual deployment, medical simulation, transcatheter aortic valve implantation.

F. Bello and S. Cotin (Eds.): ISBMS 2014, LNCS 8789, pp. 95–105, 2014.
© Springer International Publishing Switzerland 2014

1   Introduction

Cardiovascular disease has retained its position as the leading cause of deaths in the industrial world for several decades. A growing number of patients at risk can be directly linked to a growing elderly patient population [3]. In parallel, the number of patients with heart valve disease that require medical treatment is increasing. Older patients commonly face greater risk of complications during conventional surgery as shown for patients with severe aortic stenosis [9]. Therefore, it is essential to investigate novel procedures and devices related to cardiovascular treatment for this rapidly growing group of patients.

1.1   Medical Background

Aortic stenosis (AS) is a degenerative disease of the heart valve regulating the blood flow from the left ventricle to the aorta. Underlying pathology is a calcification of the aortic valve leaflets leading to a reduction of the size of the orifice of the valve resulting in an increased blood flow velocity and pressure gradient over the valve. For compensation, heart muscle cells enlarge, leading to hypertrophy of the heart and consequently to heart failure. The recommended therapy is surgical aortic valve replacement performed through invasive open-heart intervention. The outcome has shown good long-term results [14] for elective valve replacement, but is demands longer recovery time.

Transcatheter aortic valve implantation (TAVI) is a minimally invasive procedure for patients with severe aortic stenosis, who are commonly unfit for open-heart surgery. In a TAVI procedure self-expandable or balloon-expandable stents are implanted through a transapical or transfemoral approach inside the existing aortic valve position. The crimped stent is inserted into the vascular system through a catheter and led along a previously positioned guide wire. It is then expanded within the aortic root to push existing stiffened leaflets against the wall and delivers a valve prosthesis previously sewed into the device as shown in Figure 1. Each stent is available in different sizes to adapt to individual patient geometry around the aortic root.

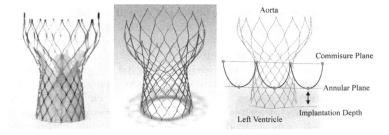

Fig. 1. *Left:* Medtronic CoreValve with porcine leaflets. *Center:* Model of the stent without leaflets. *Right:* The CoreValve has to be implanted with its base skirt in contact with the stiff annulus section of the aortic root.

1.2   Motivation

TAVI is a promising alternative to conventional open-heart surgery, but introduces limitations on accurate interactive control of the implanted prosthesis [16]. The initial decision regarding stent model and size for a specific patient is currently based on anatomical measurements extracted from a pre-interventional CT scan in combination with manufacturer recommendation and personal experience of the surgeon and interventionalist. Our workflow is designed to assist and guide this process by providing additional information about stent expansion under consideration of leaflet calcification.

Optimal stent implantation is difficult to achieve and requires extensive training as navigation during the procedure is based on a projected 2D view as shown in Figure 2. Additional surgery planning and guidance utilities are required to improve predictions regarding stent movement and deployment behavior. Finding the optimal initial placement of the prosthesis in relation to the aortic root has not been investigated sufficiently by means of FE analysis to the best of our knowledge. However, implantation depth is an important variable, which is used to measure success of the procedure as it can be monitored during the intervention [16].

1.3   Objective

Our goal is therefore to not only use simulation results for TAVI planning, but extend their use towards procedure guidance. In particular, we develop a novel workflow to introduce simulation results to a live fluoroscopic video stream that is aiming at providing predictive support during the challenging task of stent navigation. Our two main objectives are (1) to provide an enhanced workflow for TAVI planning, where simulation of the procedure can support the intervention preparation and (2) to use the results of the simulation during intervention, where it can provide additional guidance and assistance to the physician.

1.4   Related Work

Biomechanical simulation and modeling of stenting procedures has been used intensively to study stent deployment and design [7,13]. Predictive TAVI simulation has consistently received attention in literature since the procedure has been introduced, but often lacked solid validation procedures. Initially, balloon-expandable stents (e.g. Edwards Sapien) were analyzed by [1,2] and others, discussing the effect of different implant positioning, but not taking calcifications into account. [18] also focuses on balloon-expandable stents, but introduces a more complete anatomical model including basic calcifications. Self-expanding stent analysis was performed in [17], but neither applying to patient-specific models, nor analysing variable implantation depth or boundary conditions. Tracking the annulus plane and stent location in fluoroscopic images was previously implemented in [11] without considering prediction. Post-interventional deformation analysis of implanted stents has been performed to analyze force

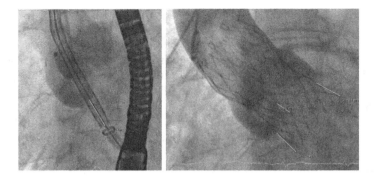

Fig. 2. Self-expanding stent and aortic root after application of contrast agents as seen during TAVI in a live fluoroscopic video image. An implantation depth of 16.8 mm was measured before deployment *(left)* and 14.0 mm after full expansion *(right)*, which corresponds to an almost optimal placement.

distributions related to deployed stents [5,8]. We have previously shown patient-specific simulation of TAVI [15] for procedure prediction. In this work we present a method to increase the degrees of freedom in relation to stent movement and study the effects of different initial deployment depths to provide a prediction framework for TAVI adapted to information available during the intervention.

2   Method

Existing TAVI simulation environments consider the implantation depth during simulation as static, and boundary conditions of the stent are defined to maintain a specific implantation depth through constraints. This implies that the stent does not move along the direction of the aorta during expansion. In reality, during stent deployment the stent is remotely controlled along this axis by a physician through a guide wire and the deployment device. The exact characteristics of the interaction with the stent through this wire depends on the surgeon's experience and can hardly be predicted. Furthermore, the stent is commonly assumed to be rotated perpendicular to the annulus plane and centered between the aortic leaflets [2], which is not generally true for all procedures as seen in Figure 2. The guide wire is commonly bent along the aortic arch and reaches into the ventricle, where it collides with the wall and results in an off-centered shift and rotation around the annular plane.

To analyze the effect of a less rigid virtual stent deployment, we propose a method that involves connecting the guide wire to the stent and therefore imposes no direct positional constraints onto the stent. Instead, its movement is restricted indirectly through the wire, which provides sufficient flexibility and allows changes in implantation depth to occur during the simulation. Our framework implements the following steps:

1. Patient-specific anatomical modeling.
2. Multi-material finite element model generation.
3. Guided stent insertion towards the annular plane.
4. Multiple stent deployment at different implantation depths.
5. Verification of expanded stent positioning.

2.1   Finite Element Model

The anatomical model for finite element mesh generation is segmented automatically from contrast enhanced CT images using a clinical prototype application (Philips HeartNavigator), which implicitly defines a number of landmark locations used to identify the annular plane, commissure plane, leaflet center and coronary arteries (Figure 3). The landmark positions are found automatically by the application. However, all landmarks have been re-defined by an expert clinician manually to increase model accuracy. Calcification within the aortic root region are segmented using a threshold of 600 HU and are classified into individual components using k-Means clustering. A centerline is generated through Voronoi skeletonization[1] and a parametric geometry model is then constructed using the commissure ($C_{1,2,3}$) and annulus ($A_{1,2,3}$) landmarks, which can be fit to the segmented data using least squares. The model, shown in Figure 3, can implicitly differentiate anatomical regions of interest, such as aorta, sinus, leaflets or ventricle for initialization of different material parameters. It is meshed using approximately 40,000 shell elements for aortic root and ventricle. Volumetric elements are used for calcifications around the leaflets.

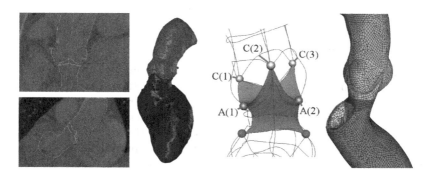

Fig. 3. CT image data are used to segment a patient-specific anatomical representation of the aortic root region. A parametric model is built based on a centerline, commissure ($C_{1,2,3}$), annulus ($A_{1,2,3}$) and leaflet landmarks.

As image based material thickness T measurement for leaflets and aortic root are depending on sufficient image quality and resolution, these and additional material parameters for biomedical simulation have to be estimated.

---

[1] Vascular Modeling Toolkit http://www.vmtk.org

Based on [12] we determine average aortic and leaflet thickness based on age groups and anatomical findings, which could be further refined in terms of, e.g., gender and ethnicity. For this study a linear elastic material model is applied with the following elastic moduli found in the literature [18,6]: Aorta: $E = 2.0\,\mathrm{MPa}$, $T = 2.5\,\mathrm{mm}$; Aortic sinus: $E = 1.8\,\mathrm{MPa}$, $T = 2.0\,\mathrm{mm}$; Annulus crown: $E = 10.0\,\mathrm{MPa}$, $T = 2.0\,\mathrm{mm}$; Aortic leaflets: $E = 1.2\,\mathrm{MPa}$, $T = 0.8\,\mathrm{mm}$; Calcification: $E = 60\,\mathrm{MPa}$, Ventricle: $E = 120.0\,\mathrm{kPa}$, $T = 11.0\,\mathrm{mm}$. The material density is defined as $1.2\,\mathrm{g/cm^3}$. The Poisson ratio of nearly incompressible solids is defined as 0.45. The friction coefficient between the tissue and stent was defined at: $\rho = 0.1\,\mu$, while the damping coefficient was set to 0.1 to stabilize the contact intensive simulation.

The self-expanding Medtronic CoreValve (Medtronic Inc., Minneapolis, MN, USA) stent was measured by acquiring micro-CT scan data [4]. It was modeled in three sizes $(26, 29, 31)$ using Timoshenko-based beam-elements [7], which are connected at their intersection points through adaptive constraints as described by [8]. Each of the 30 strings consists of 252 linear beam elements, resulting in 7560 elements for the complete stent. A linear material model (Nitinol Young's modulus: $E = 58\,\mathrm{GPa}$, Poisson number: 0.33) was shown to sufficiently represent stent deformations within the range of application of final equilibrium.

The guide wire (Steel Young's modulus: $E = 180\,\mathrm{GPa}$, Poisson number: 0.33) is initialized using 700 beam elements covering a length of $140\,\mathrm{mm}$ with a circular profile at a diameter of $0.9\,\mathrm{mm}$. Its length can be adapted to match a previously generated centerline through aorta and ventricle. The wire is linked to the stent through virtual reference points along the centerline of the stent and 15 virtual connector elements at multiple levels (Figure 4). It is constrained at the other end inside the aortic arch preventing any translational displacements.

2.2   Stent Placement Simulation

The crimping process of the stent is simulated by applying an inbound radial displacement to all nodes of the stent model, pulling them towards the centerline of the stent. The geometric tissue model is constrained at the end of the segmented aortic arch and around the lower left ventricle. A guide wire is initialized along the centerline, fixated in position only at the side of the aortic arch. The stent sheath is positioned along this wire at the opposing end. The implantation depth is measured between the annular plane and the base of the stent furthest inside the left ventricle. After stent placement the stent is constrained through the guide wire and fixated at the connecting hinge point as shown in Figure 4. The guide wire is constrained in translation at the area, where it exits the segmented part of the aorta and can otherwise move freely within the aortic root. Contact constraints are defined using the penalty method without friction between guide wire and the tissue surface.

During the next phase of stent deployment the crimping constraints are removed successively starting at the lower end of the CoreValve as if the sheath was pulled off towards the guide wire. Each of the 11 stent level is released individually. Contact computation is based on the penalty method with hard contact

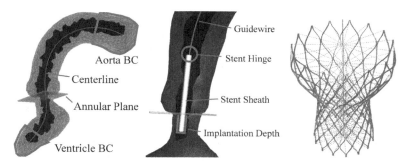

Fig. 4. View inside a segmented mesh, showing the annular plane and extracted centerline. The FE mesh will be fixated within the outlined boundary condition (BC) regions. The stent is constrained to the guide wire and positioned along the centerline (crimped inside a sheath).

along the normal direction and includes self-contact within leaflets, calcification and aortic root tissue. This formulation applies equal and opposite contact forces to nodes inside direct penetration areas. The penalty stiffness matrix K_p shown in Equation 1 is calculated every iteration and added to the stiffness matrix K_c to compute the panilty force F for the displacement u.

$$[K_c + K_p]u = F \tag{1}$$

Within the simulation we observe contact forces, strain and stress of tissue and stent, but we are particularly interested in overall changes in implantation depth relative to the annular plane.

3   Experimental Results

For this study a widely accepted commercial solver (Dassault Systems Simulia Abaqus 6.13) was used for explicit finite element simulation, where kinematic conditions are calculated in an incremental approach to solve for a dynamic equilibrium between mass matrix M, nodal accelerations $\ddot{u}$ and the nodal forces, defined by externally applied forces P and internal element forces I. The calculation of accelerations the the beginning of the current time increment are shown in Equation 2, which is integrated explicity through time t (Equation 3).

$$[M]\ddot{u} = P - I$$
$$\ddot{u}|_{(t)} = [M]^{-1} \cdot (P - I)|_{(t)} \tag{2}$$

$$\dot{u}_{(t+\Delta t/2)} = \dot{u}_{(t-\Delta t/2)} + \frac{(\Delta t_{(t+\Delta t)} + \Delta t_{(t)})}{2}\ddot{u}_t$$
$$u_{(t+\Delta t)} = u_{(t)} + \Delta t_{(t+\Delta t)}\dot{u}_{(t+\Delta t/2)} \tag{3}$$

The actual deployment accuracy has been measured through evaluation with CoreValve models segmented in six post-interventional CT scans. The circumferential difference at each level was measured to determine a rotation invariant error metric. For these experiments the implantation depth had been fixed to ensure a match between in-silico results and the actually performed intervention. An average error of 4.0 % ($\pm$2.5 %) was measured within this small sample volume. We are currently expanding the number of available datasets for wider evaluation.

The experiments for computational stent placement in TAVI involve six patient datasets and have been performed on a high-performance computer (CentOS), allocating 16 AMD Opteron (2.2 GHz) CPU cores and 16 GB RAM for each job submission. Stent deployment has been initiated at an implantation depth of $1.0, 4.0, 8.0, 12.0$ and 16.0 mm below the annular plane, with an optimal relative depth deployment specified between 5.0 to 10.0 mm [10]. The adaptive simulation time-step has a lower bound of $1e^{-6}s$ and is initiated at $0.1\ s$. Depending on the simulated case the processing time for this configuration is between 4.5 to 8 hours. After an equilibrium was reached the centerpoint of the base CoreValve level is used to determine the new implantation depth.

The results shown in Figure 5 constitute a difference in implantation depth before and after stent deployment. There is a slight tendency for the implantation depth to increase, i.e. for the stent to migrate towards the ventricle. This might partially be explained with the lower end of the stent expanding first and resolving contact within the area below the aortic leaflets, underneath the annulus plane. In our experiments, this effect appears less prominent for larger initial implantation depths, when the stent is placed deeper inside the ventricle. The overall mean change in implantation depth before and after full deployment is 0.94 mm ($\pm$1.62 mm) towards the ventricle.

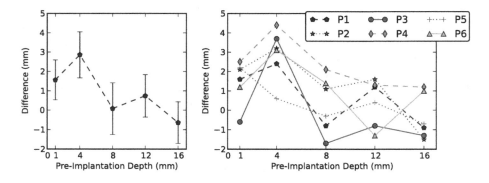

Fig. 5. *Left:* Mean difference in implantation depth before and after deployment. *Right:* Change in implantation depth for each patient (positive: towards ventricle, negative: towards aorta).

Stent and aortic root tracking on a 2D projection view based on template matching has been used by [11] to introduce additional guidance visualizations as overlays to the fluoroscopic video. The algorithm extracts the annular plane as well as the stent location and uses the known transformation matrix of the imaging device to project an aortic root into the scene. As shown before, we can simulate a number of likely locations for stent deployment and extend this method to preview full stent deployment as the actual implantation depth can be extracted from the image. We can further make a prognosis in which direction the stent is likely to move during deployment, using pre-computed simulation results. Thus, our framework allows to introduce a deployment preview, which was not possible with previous implementations.

4   Conclusion

We proposed an automated workflow for patient-specific computational TAVI stent-placement simulation under realistic conditions found during surgery. By using a TAVI stent deployment approach following closely the clinical practice, we created a tool that may assists surgeons to predict the movement of implants from any initial position. The less restraint boundary conditions of our model allow prediction of complex dynamics during stent deployment, which was not possible previously. The results show a necessity of the added degree of freedom to determine changes in stent implantation depth. Even though, our experiments do not take into account the live interaction a physician could perform, they provide recommendations for adjusting stent placement prior to deployment as our prediction takes into account different initial positions relative to the annular plane. The underlying finite element simulation has been validated using pre- and post-operative CT scans. Our workflow has been demonstrated experimentally on six patient cases. Future work includes acquisition of additional datasets. We will also extend our sensitivity analysis to validate the effects of changes in tissue thickness and elasticity material parameters. Further, it would be relevant to analyze the influence of blood pressure and ventricular contraction. The guide wire initialization could also be improved by simulating the actual insertion procedure prior to stent positioning.

Our results show that the initial placement of the prosthesis is critical to its behavior during deployment, which is an expected result. However, it was previously not possible to predict deployment shifts during surgery, where patient-specific anatomy and calcification as well as the real world stent placement are taken into account. We were able to show that different stent initialization positions can result in large differences in stent drift and therefore the final implantation depth.

Acknowledgments. This work was supported by the Swiss National Science Foundation (SNSF grant no. CR32I3_135044) and the Swiss Heart Foundation.

References

1. Auricchio, F., Conti, M., Morganti, S., Reali, A.: Simulation of transcatheter aortic valve implantation: a patient-specific finite element approach. Computer Methods in Biomechanics and Biomedical Engineering 17(12), 37–41 (2013)
2. Capelli, C., Bosi, G.M., Cerri, E., Nordmeyer, J., Odenwald, T., Bonhoeffer, P., Migliavacca, F., Taylor, A.M., Schievano, S.: Patient-specific simulations of transcatheter aortic valve stent implantation. Medical & Biological Engineering & Computing 50(2), 183–192 (2012)
3. Finegold, J.A., Asaria, P., Francis, D.P.: Mortality from ischaemic heart disease by country, region, and age: statistics from World Health Organisation and United Nations. International Journal of Cardiology 168(2), 934–945 (2013)
4. Gessat, M., Altwegg, L., Frauenfelder, T., Plass, A., Falk, V.: Cubic hermite bezier spline based reconstruction of implanted aortic valve stents from CT images. In: Conf. Proc. IEEE Eng. Med. Biol. Soc. (2011)
5. Gessat, M., Hopf, R., Pollok, T., Russ, C., Frauenfelder, T., Sündermann, S.H., Hirsch, S., Mazza, E., Székely, G., Falk, V.: Image-Based Mechanical Analysis of Stent Deformation. Concept and Exemplary Implementation for TAVI Stents. IEEE Transactions on Biomedical Engineering 60(1), 4–15 (2014)
6. Grbic, S., Mansi, T., Ionasec, R., Voigt, I., Houle, H., John, M., Schoebinger, M., Navab, N., Comaniciu, D.: Image-Based Computational Models for TAVI Planning: From CT Images to Implant Deployment. In: Mori, K., Sakuma, I., Sato, Y., Barillot, C., Navab, N. (eds.) MICCAI 2013, Part II. LNCS, vol. 8150, pp. 395–402. Springer, Heidelberg (2013)
7. Hall, G.J., Kasper, E.P.: Comparison of element technologies for modeling stent expansion. J. Biomechan. Eng. 128(5), 751–756 (2006)
8. Hopf, R., Gessat, M., Falk, V., Mazza, E.: Reconstruction of Stent Induced Loading Forces on the Aortic Valve Complex. In: Demirci, S., Unal, G., Lee, S.-L., Radeva, P. (eds.) Proceedings of MICCAI-Stent 2012, Nice, France, pp. 104–111 (2012)
9. Iung, B., Cachier, A., Baron, G., Messika-Zeitoun, D., Delahaye, F., Tornos, P., Gohlke-Bärwolf, C., Boersma, E., Ravaud, P., Vahanian, A.: Decision-making in elderly patients with severe aortic stenosis: why are so many denied surgery? European Heart Journal 26(24), 2714–2720 (2005)
10. Jilaihawi, H., Chin, D., Spyt, T., Jeilan, M., Vasa-Nicotera, M., Bence, J., Logtens, E., Kovac, J.: Prosthesis-patient mismatch after transcatheter aortic valve implantation with the Medtronic-Corevalve bioprosthesis. European Heart Journal 31(7), 857–864 (2010)
11. Karar, M.E., John, M., Holzhey, D., Falk, V., Mohr, F.-W., Burgert, O.: Model-updated image-guided minimally invasive off-pump transcatheter aortic valve implantation. In: Fichtinger, G., Martel, A., Peters, T. (eds.) MICCAI 2011, Part I. LNCS, vol. 6891, pp. 275–282. Springer, Heidelberg (2011)
12. Malayeri, A.A., Natori, S., Bahrami, H., Bertoni, A.G., Kronmal, R., Lima, J.A.A.C., Bluemke, D.A.: Relation of aortic wall thickness and distensibility to cardiovascular risk factors (from the Multi-Ethnic Study of Atherosclerosis [MESA]). The American Journal of Cardiology 102(4), 491–496 (2008)
13. Morlacchi, S., Chiastra, C., Gastaldi, D., Pennati, G., Dubini, G., Migliavacca, F.: Sequential Structural and Fluid Dynamic Numerical Simulations of a Stented Bifurcated Coronary Artery. J. Biomechan. Eng. 133(12), 121010 (2011)
14. Rosenhek, R., Zilberszac, R., Schemper, M., Czerny, M., Mundigler, G., Graf, S., Bergler-Klein, J., Grimm, M., Gabriel, H., Maurer, G.: Natural history of very severe aortic stenosis. Circulation 121(1), 151–156 (2010)

15. Russ, C., Hopf, R., Hirsch, S., Sündermann, S.H., Falk, V., Székely, G., Gessat, M.: Simulation of Transcatheter Aortic Valve Implantation under Consideration of Leaflet Calcification. In: 35th Annual International Conference of the IEEE Engineering in Medicine and Biology Society (2013)
16. Toggweiler, S., Webb, J.G.: Challenges in transcatheter aortic valve implantation. Swiss Med. Wkly. (December 2012)
17. Tzamtzis, S., Viquerat, J., Yap, J., Mullen, M.J., Burriesci, G.: Numerical analysis of the radial force produced by the Medtronic-CoreValve and Edwards-SAPIEN after transcatheter aortic valve implantation (TAVI). Medical Engineering & Physics 35(1), 125–130 (2013)
18. Wang, Q., Kodali, S., Primiano, C., Sun, W.: Simulations of transcatheter aortic valve implantation: implications for aortic root rupture. Biomechanics and Modeling in Mechanobiology (April 2014)

Testbed for Assessing the Accuracy of Interventional Radiology Simulations

Mario Sanz-Lopez[1], Jeremie Dequidt[2], Erwan Kerrien[1,3], Christian Duriez[1,2], Marie-Odile Berger[1,3], and Stéphane Cotin[1,2]

[1] INRIA, France
[2] University of Lille, France
[3] University of Nancy, France

Abstract. The design of virtual reality simulators, and more specifically those dedicated to surgery training, implies to take into account numerous constraints so that simulators look realistic to trainees and train proper skills for surgical procedures. Among those constraints, the accuracy of the biophysical models remains a very hot topic since parameter estimation and experimental validation often rely on invasive protocols that are obviously not suited for living beings. In the context of Interventional Radiology the procedures involve the navigation of surgical catheter tools inside the vascular network where many contacts, sliding and friction phenomena occur. The simulation of these procedures require complex interaction models between the tools and the blood vessels for which there is no ground truth data available for parametrization and validation. This paper introduces an experimental testbed to address this issue: acquisition devices as well as a data-processing algorithms are used to record the motion of interventional radiology tools in a silicon phantom representing a vascular network. Accuracy and high acquisition rates are the key features of this testbed as it enables to capture dynamic friction of non-smooth dynamics and because it could provide extensive data to improve the accuracy of the mechanical model of the tools and the interaction model between the tools and the blood vessel.

1 Introduction

Interventional radiology procedures rely on the navigation of a catheter inside the blood vessels through fluoroscopic image guidance. When the tip of the catheter is close to the location of the disease, dedicated tools (coils, stents, balloon...) can be deployed or drugs can be delivered in order to diagnose or treat the patient. To control the motion of a catheter within the vascular network, the radiologist can only push, pull or twist the proximal end of the device. She/he also can insert a stiffer device, named *guidewire*, that will help reaching specific locations. Since such devices are constrained inside the patient's vasculature, the motion of the devices results from the displacement inputs of the radiologist as well as contact forces between the devices and the arterial walls. These contact forces are particularly important since the motion of the devices is greatly

F. Bello and S. Cotin (Eds.): ISBMS 2014, LNCS 8789, pp. 106–111, 2014.

affected by dynamic effects such as friction or sliding which can significantly change the trajectory followed by the devices. Even if the friction between the arterial wall and the tools may be negligible in certain circumstances (for instance slow motion, simple path to reach the targeted area), its effects are significant in tortuous vasculature or aneurysms. The figure 1 illustrates that friction implies high frequency motion for the deployment of a coil (thin coated platinum thread) in an aneurysm. Therefore, accurate models for the tools and for the interactions between tools and the blood vessel are mandatory. However, the quantitative evaluation of the accuracy is hard to perform due to the complexity of the environment: while imaging techniques allow to capture the geometry of the anatomical structures, the bio-mechanical parameters are hard to measure / estimate. Some previous work has proposed strategies to estimate the accuracy of their simulations: Dequidt *et. al.* [4] use 3D rotational angiography to evaluate the simulation of a coil deployment in a collision-free environment. Luboz *et. al.* [5] compare the position of simulated guidewires and actual guidewires inserted in a vascular phantom. The comparison is however performed on equilibrium positions and on a almost linear trajectory which reduces the influence of friction. Finally, Alderliesten *et. al* [1] have conducted a thorough quantitative experimental validation of a quasi-static model of guidewires: 3D comparison through 3DRA is performed with several deployment scenarios. However the main limitation of the approach is that only static friction is considered and the friction parameter is empirically set. Even if the friction between the arterial wall and the tools may be negligible in certain circumstances, its effects are significant in tortuous vasculature or aneurysms even if special care is taken by the radiologist with slow and delicate manipulations. For instance, friction implies high frequency motion for the deployment of a coil (thin coated platinum thread) in an aneurysm. To capture this effect on navigation scenarios, our objective is to design a testbed with software support that is able to record dynamic and non-smooth phenomena (i.e high speed changes). This paper describes an experimental testbed suited for navigation scenarios that include motion tracking of the tool but also image tracking through high-speed stereoscopic cameras.

2   Validation Testbed

The testbed proposed in this paper is designed to provide ground-truth data to compare against simulation. To do so, the testbed is composed of two components: (1) a tracking device is built to accurately track the input motion of the surgeon navigating the radiology tools, (2) high-speed stereoscopic cameras coupled with segmentation and reconstruction provide the position of the tool in a vascular silicon phantom.

Motion Tracking: We have designed a new motion tracking interface, based on an optical sensor from a mouse that detects motion with a 1200 dpi resolution, a standard HID USB interface and a radio wireless connexion to transmit the raw data without cluttering the operating field. A new package has been built

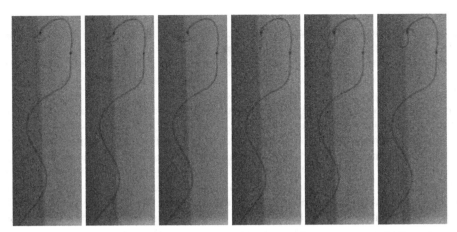

Fig. 1. Recorded sequence of a coil deployed at low and constant speed where image acquisition is performed at 15 fps through 3D Rotational Angiography. The sequence shows the effect of friction and then sliding as the coil completely changes its configuration between frames 3 and 4 (67 msecs between each frame).

for the sensor allowing an easy handle, standard surgical attach systems on both sides and removable tunnels adapted to each size of catheters and coils. The main issues with the use of a mouse sensor relies in sensing the cylinder shape of the catheter instead of a flat surface, the placement of the circuit in a different embodiment, as well as the existence of a wireless, potentially noisy link rise questions about its reliability. Moreover, the data provided is differential, which could lead to a cumulative input error. Thus, a precision test has been conducted considering both local (resolution) and absolute accuracy using a calibrated test bench.

This test bench is composed of a motorized spool that pulls over one meter of bare electric wire, with a negligible elongation factor and a constant diameter of

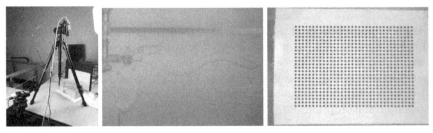

Fig. 2. (left) Experimental setup, showing the top and side camera, observing the silicon vascular phantom; (middle) image of the phantom by the top camera; (right) calibration pattern used to compute the equation of the refractive plane (inter-bullet distance=2 mm)

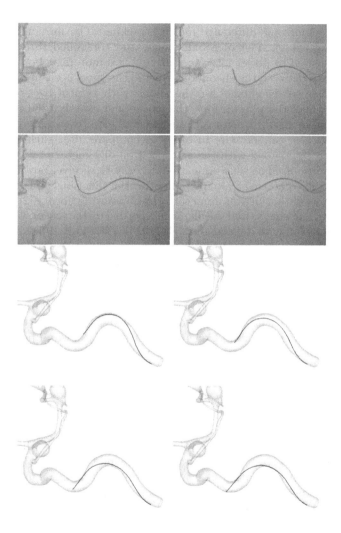

Fig. 3. Stick and slip transition: (top) top camera raw view, (bottom) reconstruction. Only one out of two images of the original sequence were shown for a better readability. A top speed of 485.13 mm/s was recorded at the start of the sequence.

1.54mm (close to interventional radiology tools). While the electric wire diameter is greater that the actual interventional radiology tools, it was chosen due to its non-elongation property and we use dedicated diameter-dependent removable tunnels that allow to place any thin wire in the focal plane of the sensor in order to maximize the motion tracking. Moreover the wire is not insulated as it does not provide a significant difference for the tracking. The motion of the spool is tracked using the encoder of a ball mouse and controlled through a micro-controller connected to power interfaces driving two motors at both sides of the cylinder; the grip between the encoder wheel and the spool is ensured by choosing a silicone rubber solid cylinder for the spool, with typical friction coefficient larger than 1. Under these conditions, the mouse sent exactly 1260 counts, which gave a final system absolute error of $\pm396\mu m$, that can be considered negligible. A landmark was set on the wire in order to compare after each test that the mark was perfectly aligned with the goal position of 1 meter of wire pulled.

The laser mouse used as sensor makes data easy to receive, but it needs to be fetched before the Operating System applies transfer functions and de-naturalizes the raw recovered data, this was done with the libpointing library [2]. 6 tests were conducted at various speeds and the distance recorded by the sensor was stored. For 1 meter of wire pulled with a speed of (0.0487, 0.0967, 0.145, 0.192, 0.242 and 0.290 m/s) the sensed distance was (1051.200, 1006.284, 1009.480, 1064.662, 1060.026 and 991.4255 mm) which leads to an average error of 3.3%.

High-Speed Stereoscopic Tracking: The system was centered on two JAI-Pulnix TM-6740CL cameras configured to acquire 640x480 grayscale images at 198 frames per second. Synchronization was performed thanks to a trigger (CC320 Machine Vision Trigger Timing Controller, Gardasoft). Two Super Cool-Lite 9 (interFit Lighting) projectors were used to reduce the flickering observed with neon lights. Both cameras were placed to respectively look down and sideways (see Fig. 2, left). The stereo pair of cameras were calibrated using chessboard detection and calibration by OpenCV.

The vasculature was given by a phantom made of silicon (right internal carotid artery rigid model with three aneurysms, by Elastrat). Silicon is transparent (see Fig. 2, middle) but induces refraction that significantly perturbs the epipolar geometry of the stereoscopic cameras and forbids using traditional 3D reconstruction based on the triangulation of visual rays [3]. The plane equations of both refractive faces were given by computing the associated homography thanks to a calibration pattern and OpenCV (see Fig. 2, right).

We used a straight shaped microcatheter with an outside diameter of 1.7 French (Headway^{TM}17 regular Microcatheter, Microvention) that was only pushed into the phantom. The microcatheter could automatically be segmented after subtracting a catheter-free image. Refractive stereoscopic reconstruction was thereafter applied followed by an RBF (Radial Basis Functions) smoothing of the curve. The reconstruction error was estimated to be below 0.05 mm on 1000 images of a motionless catheter. A 3D model of the phantom vasculature was segmented from 3DRA data using [6,7] and manually registered with the images of both cameras. In order to estimate the imprecision of the reconstruction, we measured the average displacement of the catheter tip at the start of the sequence, while no motion was applied (1000 images): a standard deviation of [0.0075, 0.01750.0457] mm was observed along the X-Y-Z axes (Y and Z axes are respectively along the optical axes of the side and top cameras).

3   Results and Conclusion

We focused our analysis on abrupt motions. Fig. 3 gives an example of a stick and slip transitions: starting from a static position of the catheter tip, a motion amplitude of 10.1 mm was observed on 8 consecutive images, which gives an average speed of 286.86 mm/s. Over this sequence, the instantaneous speeds, as measured in-between two consecutives images, were : [7.06, 123.74, 485.13, 400.04, 388.54, 388.98, 326.56, 27.75] mm/s. A peek of almost 500 mm/s was observed at the start of the sequence.

These first results demonstrate that we have developed a testbed able to validate numerical models and simulations of the mechanical interaction between catheter and blood vessels. In future work, we will use this experimental ground truth to validate and compare existing simulation algorithms. The testbed has been built using a rigid silicon phantom as it allows to simplify the actual environment and puts the emphasis on the tool model and the interaction model between the tool and the blood vessel. Using a soft silicon phantom will be considered in a future work to validate vessel-deformation models but will require a more complex machinery (regular segmentations of the vasculature will be mandatory to capture the deformations of the soft silicon phantom).

References

1. Alderliesten, T., Konings, M., Niessen, W.: Modeling friction, intrinsic curvature, and rotation of guide wires for simulation of minimally invasive vascular interventions. IEEE Transactions on Biomedical Engineering 54(1), 29–38 (2007)
2. Casiez, G., Roussel, N.: No more bricolage!: Methods and tools to characterize, replicate and compare pointing transfer functions. In: UIST, pp. 603–614 (2011)
3. Chari, V., Sturm, P.: Multiple-view geometry of the refractive plane. In: BMVC 2009 - 20th British Machine Vision Conference, pp. 1–11 (2009)
4. Dequidt, J., Marchal, M., Duriez, C., Kerien, E., Cotin, S.: Interactive simulation of embolization coils: Modeling and experimental validation. In: Metaxas, D., Axel, L., Fichtinger, G., Székely, G. (eds.) MICCAI 2008, Part I. LNCS, vol. 5241, pp. 695–702. Springer, Heidelberg (2008)
5. Luboz, V., Zhai, J., Odetoyinbo, T., Littler, P., Gould, D., How, T., Bello, F.: Simulation of endovascular guidewire behaviour and experimental validation. Computer Methods in Biomechanics and Biomedical Engineering 14(06), 515–520 (2011)
6. Yureidini, A., Kerrien, E., Cotin, S.: Robust RANSAC-based blood vessel segmentation. In: Haynor, D.R., Ourselin, S. (eds.) SPIE Medical Imaging, vol. 8314, p. 8314M. SPIE Press (February 2012)
7. Yureidini, A., Kerrien, E., Dequidt, J., Duriez, C., Cotin, S.: Local implicit modeling of blood vessels for interactive simulation. In: Ayache, N., Delingette, H., Golland, P., Mori, K. (eds.) MICCAI 2012, Part I. LNCS, vol. 7510, pp. 553–560. Springer, Heidelberg (2012)

Simulation of Catheters and Guidewires for Cardiovascular Interventions Using an Inextensible Cosserat Rod

Przemyslaw Korzeniowski[1], Francisco Martinez-Martinez[2], Niels Hald[1], and Fernando Bello[1]

[1] Department of Surgery and Cancer,
Imperial College London
[2] Inter-University Research Institute for Bioengineering and Human Centered Technology,
Polytechnic University of Valencia

Abstract. Effective and safe performance of cardiovascular interventions requires excellent catheter / guidewire manipulation skills. These skills are mainly gained through an apprenticeship on real patients, which may not be safe or cost-effective. Computer simulation offers an alternative for core skills training. However, replicating the physical behaviour of real instruments navigated through blood vessels is a challenging task.

We use an inextensible Cosserat rod and impulse-based techniques to model virtual catheters and guidewires. This allows an efficient recreation of bending, stretching and twisting phenomena of the material in real-time. It also guarantees an immediate response to user manipulations even for long instruments. The mechanical parameters of six guidewires and three catheters were optimized with respect to their real counterparts scanned in a silicone phantom using CT.

The validation results show near sub-millimetre accuracy with an average distance error between the trajectories of the simulated and scanned instruments of 1.34mm (standard deviation: 0.95mm, RMS: 1.66mm). Our implementation requires just 0.2ms per time step to process 200 Cosserat elements on an off-the-shelf laptop, enabling simulation of 40cm long instruments at 4 kHz, thus significantly exceeding the minimum required haptic interactive rate (1 kHz).

Keywords: Catheter, Guidewire, Cosserat Rod, Medical Simulator.

1 Background

Cardiovascular diseases (CVD) are the main cause of death in the developed world [1]. Minimally invasive endovascular procedures, widely adopted in diagnosis and treatment of CVDs, improve recovery time, reduce patient trauma and health-care costs. During such procedures, endovascular clinicians insert long, thin, flexible surgical instruments – catheters and guidewires, into the patient's vascular system. Guided by medical imaging, they then navigate the catheter / guidewire pair into the coronary arteries to treat the pathology. An effective and safe performance of these

F. Bello and S. Cotin (Eds.): ISBMS 2014, LNCS 8789, pp. 112–121, 2014.
© Springer International Publishing Switzerland 2014

procedures requires excellent instrument manipulation skills, which are still mainly gained through an apprenticeship on real patients. Drawbacks of the apprenticeship model include costs, reduced training opportunities and patient-safety [2].

One possible alternative is training on computer-based, virtual reality (VR) simulators [3]. The last decade has seen growing interest in the benefits of using VR medical simulators in a range of specialties, including endovascular interventions [4-8]. Commercial VR vascular simulators such as VIST (*www.mentice.com*) or Angio Mentor (*www.simbionix.com*) have demonstrated a degree of face and content validity, but the ultimate realism is yet to be achieved [4].

The fundamental part of such simulators is an underlying mathematical model of the virtual catheter and guidewire. [5] and [6] showed the possibility to navigate a guidewire in real-time and with visually correct accuracy using a mass-spring model. Thanks to increasing computational power, solutions based on continuum mechanics approaches such as the finite element method (FEM) have become feasible. [7] introduced a non-linear deformable beam model resulting in an accurate simulation, whilst [8] simulates a guidewire, including friction, as a set of straight, non-bendable, incompressible beams with perfect torque control using a quasi-static approach.

The above solutions convincingly reproduce the bending phenomenon of the material, but either ignore the twisting deformation or recreate it in an inefficient or inaccurate way for real-time simulation. In [9], Pai proposed an alternative model based on an established theoretical framework – the Cosserat theory of elastic rods [10]. Pai's model – the Strand – realistically reproduces all the deformations of an elastic rod. However, it is limited to static deformations, which makes it difficult to include collision handling. Wen et al. [11] solved dynamic deformations of guidewires using a discrete differential geometry formulation. Spillmann and Teschner [12] proposed the CoRdE, a model also based on Cosserat theory considering dynamic effects. In [13], they improved the CoRdE model by making it inextensible. Duratti [14] applied a solution resembling the CoRde model to real-time interventional radiology simulation.

Our implementation presented here is based on the CoRdE model [12] with rod inextensibility, collision response and instrument interactions handled by our constraints framework customized for simulating catheters and guidewires. We focused on real-time performance and realistic behaviour of tools and their interaction. This enables simulation of long wires interacting with each other at haptic interactive rates, fast response to user manipulations and an easy parameterization of the mechanical properties of the instruments. We recreated six real guidewires and three catheters by optimizing their parameters with respect to their real counterparts scanned in a silicone phantom model using CT. The mechanical parameter optimization takes into account the requirements for real-time usability and stability.

The remainder of this paper is as follows: the next section gives a brief overview of the Cosserat model and presents our implementation. It also explains how the parameter optimization and evaluation were carried out. Section 3 outlines the results, whilst conclusions are presented in the final section.

2   Methods

2.1   Cosserat Rod

An elastic rod is a long and thin body able to sustain large global deformations, even if local strains are small. The Cosserat rod is a non-linear elastic rod with an oriented centreline, which enables modelling of bending and twisting deformations. A comprehensive theory is given in the book "Nonlinear problems of elasticity" by Antman [10]. Here we summarize the model as explained in [12] and [13] presenting our own implementation.

The centreline of the rod is represented by a function mapping line parameter s to a position in 3D space $\mathbf{r} = \mathbf{r}(s): [0, 1] \rightarrow \mathbb{R}^3$. In order to represent bending and twisting deformations, the concept of material frames is introduced. The material frame is an orthonormal basis $\mathbf{d}_k$, $k = 1, 2, 3$, where $\mathbf{d}_k$ are called directors. The first and second director indicate the orientation of the centreline, whereas the third one, $\mathbf{d}_3(s)$, is always adapted to the curve, i.e. parallel to the tangent $\mathbf{r}'(s)$ at the same point (see Fig. 1). From the directors we derive the rotation matrix $\mathbf{R}(s) \in \mathbb{R}^{3 \times 3}$ expressed in the implementation by a unit quaternion $\mathbf{q}$.

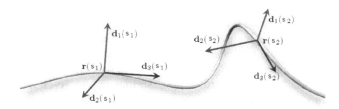

Fig. 1. The material frames adapted to the rod's centreline. With permission from [12].

By assuming that the rod is unshearable and inextensible, it is possible to set aside the strains of the centreline i.e. its stretch. Using differential geometry, from the material frames we obtain an orientational rate of change in the reference frame – the *Darboux vector* – $\mathbf{u_0} \in \mathbb{R}^3$. Its components are the areas swept by the directors when proceeding from s to $(s + \Delta s)$.

$$u_0(s) = \tfrac{1}{2}\sum_{k=1}^{3} d_k(s) \times d_k'(s) \tag{1}$$

d_k' is a partial derivative of the material frame with respect to the line parameter s. After rotating $\mathbf{u_0}$ from the reference frame into the local frame $\mathbf{u} = \mathbf{R}^T\mathbf{u_0}$, $\mathbf{u}$ relates to bending and twisting strains. Basing on the defined strain rates and assuming a linear strain-stress relationship, we derive the potential bending energy of the entire rod:

$$V_b = \tfrac{1}{2}\int_0^1 \sum_{k=1}^{3} K_k(u_k - \hat{u}_k)^2 \, ds \tag{2}$$

where $K_1 = K_2 = E\frac{\pi r^2}{4}, K_3 = G\frac{\pi r^2}{2}$, E and G are Young's and shear modulus governing the bending and torsional resistance, r is the radius of the rod's cross-section and $\hat{u}_k$ are the intrinsic bend and twist parameters. They are used to control the resting shape of the rod, for example, to model curved tips of guidewires. By minimizing V_b and by treating bending and torsion in a unified manner, we coupled the strain rates together. The twist deformation is balanced out by the bend deformation and vice versa, which results in the looping phenomenon.

2.2   Constraints

The bending energy as defined above depends solely on the configuration of the material frames. In order to couple the material frames with the centreline, we define a parallel constraint C_p (3), which aligns the third director $\mathbf{d}_3$ of the material frame with the tangent $\mathbf{r}'_i$ of the centreline.

$$C_p = \mathrm{r}' - \mathrm{d}_3 = 0 \tag{3}$$

To maintain the above constraint, the penalty method is used resulting in the following penalty energy equation:

$$E_p = \frac{1}{2}\int_0^1 K_p \left(\frac{\mathrm{r}'}{\|\mathrm{r}'\|} - \mathrm{d}_3\right) \cdot \left(\frac{\mathrm{r}'}{\|\mathrm{r}'\|} - \mathrm{d}_3\right) \mathrm{d}s \tag{4}$$

where K_p is a numerical, experimental spring constant that depends on the simulated material, and $\|\mathrm{r}'\|$ is the element length. The advantage of the penalty method is its simplicity and fast computation times. It acts as an additional force on the mass-points and torque on material frames. The disadvantage is the possible loss of accuracy, as the constraint may not be always exactly satisfied. In addition, higher values of K_p, needed for stiffer materials simulation, require smaller time step. Nevertheless, as catheters and guidewires are, by definition, flexible instruments, we find the penalty method accurate enough to simulate them. The second constraint is a quaternion unit constrain. It is enforced by coordinate projection, i.e. renormalizing the quaternions at each simulation step.

In the previous section we assumed the rod to be inextensible, thus we didn't derive an energy equation governing the stretch of the centreline. For an extensible rod, this can be found in [12]. Instead, and similarly to [13], in order to maintain the desired rest length between the centreline's mass-points (distance constraint), we employed *Lagrange multipliers* using the $\mathbf{JM^{-1}J^T}$ projection. Additionally, we used constraints to model other aspects of the simulation. Binding constraints, which are effectively distance constraints with a desired rest length set to zero, keep the guidewire within the catheter. Contact constraints are responsible for handling collisions and keeping the instruments inside vessels. However, instead of using acceleration-level constraints as suggested in [13], we apply impulse-based methods at the velocity-level. In other words, to preserve the constraint, we compute a sudden change in momentum – an impulse – which, input into the equations of motion, eliminates velocities that violate the constraints. These are called "hard constraints", as opposed to "soft constraints" enforced using the penalty method.

At first, we solved the constraints iteratively (*Projected Gauss Seidel* or *Sequential Impulses*). However, under excessive forces, the distance constraints tended to visibly compress or stretch. Increasing the number of iterations helped to reduce this effect, but significantly affected the performance. To tackle this, we combined global and iterative methods. The distance constraints are implemented in a global matrix form and are all solved together at once. As they form a chain, i.e. the first mass-point connects to the second one, the second one to the third one and so on, we are able to exploit this specific structure to significantly speed up the matrix operations. Moreover, the resulting $JM^{-1}J^T$ matrix governing only distance constraints is a symmetric tridiagonal matrix that can be efficiently solved in linear time. After applying the distance constraints, we compute binding and collision constraints and apply them locally. The described procedure can be repeated in order to "harden" the constraints, but a single iteration is usually sufficient for our application.

In sum, after receiving the user input from the haptic device, we differentiate the bending energy V_b and the penalty energy E_p with respect to the coordinates to get the stresses, i.e. restitution forces and torques which accelerate the centreline points and material frames to equilibrium. After integrating these forces, we constrain the resulting velocities and integrate them into new positions and orientations.

To compare, in the paper on inextensible version of CoRdE [13], the authors assemble a much larger $JM^{-1}J^T$ matrix governing parallel and unit quaternion constraints (distance constraints are handled implicitly). This requires handling 4 extra degrees of freedom of the quaternions and 4 times more Jacobian rows. It eliminates the penalty method and the problems related with it mentioned earlier, but involves much more computation and results in a band matrix with a bandwidth 9 to solve. Obviously, [13] does not consider the interaction between the catheter and guidewire (binding constraints) as this is specific to our application. Adding them would further increase the bandwidth. Moreover, acceleration level constraints are more computationally intensive than velocity constraints as they require calculating the second derivative of the Jacobian.

The approach by Wen et al. [11] [15], based on [16], uses parallel transport to compute elastic forces, which also eliminates the penalty method in parallel constraints. The material torsion is not computed for each centerline segment, but only at the boundary segments. This results in a faster response to torsion, but, according to [17], a quasi-static update of material frames neglects the moment of inertia of the rod's cross-section. The inextensibility (distance constraints) is handled similarly as in the original CoRdE, by using the penalty method [12], whereas collision response is implemented using impulses [15]. In the results section we briefly compare the error and performance of our approach to that of Wen et al.

2.3   Evaluation Methodology

In total, nine commonly used instruments made by Cook Medical Inc., Boston Scientific and Terumo Corp. were chosen for the simulation: three access guidewires (Cook Fixed Core Straight, Cook T-J-curved and Boston Bentson); a selection guidewire (Terumo Stiff Angled); two exchange guidewires (Boston Amplatz Super Stiff and

Cook Rosen-Curved); and three diagnostic catheters (5F Beacon, Terumo 4Fr and Terumo 5Fr). Each instrument was inserted by the same operator at room temperature into a silicone vascular phantom (*Elastrat*). The set-up was scanned with the instruments reaching three different anatomical locations (see Fig. 2): common iliac artery bifurcation, aortic bifurcation and left renal artery origin. The scanner was a multidetector CT with a resolution of 0.53x0.53x1 mm3. The 3D geometry of the phantom and inserted instruments was reconstructed using the snake segmentation algorithm in ITK-Snap. The instrument geometries were further processed to obtain their reference centrelines as showed in Fig. 2.

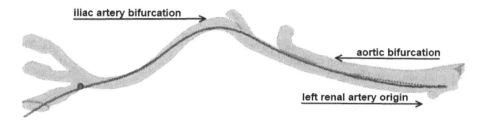

Fig. 2. Reconstructed 3D geometry of the phantom showing the centreline of the Cook J guidewire in red and the simulated instrument centreline in green

The virtual instruments were then automatically inserted into the virtual phantom. To ensure that the obtained parameters are valid for the required real-time performance, the physics simulation rate was capped at a steady 4 kHz, significantly exceeding the minimum required haptic interactive rate. Matlab's Genetic Algorithm Toolbox [18] was used to find optimal parameters of the Cosserat rod model. The population size was set to 50 individuals and the algorithm stopped after 50 generations. The optimization took close to 2500 simulations per rod to converge (approx. 2 days). Each simulation involved automated insertion of the rod to the two first error measurement locations: iliac artery bifurcation and aortic bifurcation. The optimization consisted of the minimization of an error function calculated as a root mean square (RMS) distance between the simulated mass-points and nearest points on the reference centreline and dependent on the following model parameters: Young modulus (*E*), radius (*r*), density (*d*), spring constant of the parallel constraint (*K_p*) and the ratio between the Young modulus of the tip and the shaft (*α*). The error function was an average RMS distance between simulated and reference centrelines (5).

$$f\left(E_b, r, d, K_p, \alpha\right) = \sqrt{\frac{1}{N}\sum_{i=1}^{N}(\text{minDist}(p_i^s))^2} \tag{5}$$

where **N** is the number of simulated mass-points inserted into the phantom and the function $\text{minDist}(p_i^s)$ returns the nearest Euclidean distance to the reference centreline from the position of the i-th simulated mass-point – p_i^s.

Due to availability of just one scanned dataset, the optimal parameters were obtained using the average RMS error value at the iliac artery bifurcation and aortic bifurcation. Next, they were validated at the deepest insertion reaching the left renal artery origin. In the case of instruments with curved tips, the orientation aligning the virtual and reference tips in the same plane was manually obtained before parameter optimization, and used during automated insertion to rotate the virtual instrument tip between the three anatomical locations to the correct orientation.

3   Results

Table 1 summarizes the validated RMS error for each simulated instrument only at the deepest insertion (left renal artery) and the RMS error, Euclidean distance error, standard deviation and maximum Euclidean distance (mm), all averaged using all three insertion depths, between the simulated instrument and the nearest corresponding points on the reference centreline (also averaged). For two of the instruments (Rosen, Beacon) the accuracy obtained was at the sub-millimetre level and for Terumo ST4 and ST5 catheters slightly above. The highest error was for the stiff Amplatz guidewire - 4.33mm. Average result at the 3rd depth was 2.28mm.

Table 1. Comparison between real and simulated instruments

	3rd depth RMS (mm)	Avg. RMS (mm)	Avg. Eucl. Dist. (mm)	Avg. Std. dev. (mm)	Avg. Max. Dist. (mm)	Eb (1×10^7)	Radius (1×10^3)	Density (1×10^5)	Kp (1×10^3)
Amplatz	4.33	2.60	1.94	1.66	6.70	3.65	8.9	8.47	1.63
Cook Str.	2.84	1.78	1.31	1.15	5.01	3.28	6.5	3.09	39.9
Bentson	2.69	2.06	1.72	1.13	5.37	9.03	7.5	88.2	4.05
Terumo Stiff	2.90	2.02	1.59	1.20	4.99	5.45	6.2	15.0	31.2
Cook J	3.63	2.09	1.62	1.30	4.88	17.6	6.8	1.33	10.2
Rosen	0.91	0.81	0.63	0.51	4.04	3.90	6.7	7.24	37.7
Beacon 5FR	0.90	1.08	0.98	0.45	2.56	3.53	7.4	36.5	30.2
Terumo ST4F	1.18	1.33	1.19	0.58	3.57	3.66	6.2	1.80	8.92
Terumo ST5F	1.14	1.21	1.07	0.55	3.43	3.54	6.3	11.4	31.7
Average	2.28	1.66	1.34	0.95	4.50				

The average distance error between the simulated and scanned instruments, measured using all three insertions, was 1.34mm (standard deviation: 0.95mm, RMS: 1.66mm). Comparing this error to a corresponding error of 2.27mm presented in [6] obtained by a mass-spring model in an identical data set we get nearly 60% accuracy increase. Our average RMS error (1.66mm) is slightly worse to the other Cosserat implementation [11] (1.25mm). However, they used a different dataset and test methodology, which may account for some of the difference in the RMS error.

The computational time per physics iteration of the virtual instrument consisting of 200 Cosserat elements (in our case, a 40cm long instrument) on an Asus N55s laptop (Win7 x64, Intel Core i7 2.2 GHz, 8GB RAM, NVidia GeForce GT 555M) is 0.2ms. Calculation of the Cosserat internal forces is the most computational intensive step, taking 0.15ms of the total physics update. The remaining 0.05ms is distributed equally between the constraint impulse generation and the semi-implicit Euler integrator. The collision detection runs in parallel, one time-step behind to the physics thread and takes, in total, also 0.2ms.

In Table 2 we compare the performance of our implementation to three other models: the original extensible CoRdE [12], inextensible modification by the same authors [13] and other recent inextensible Cosserat model [11]. Note that different test platforms were used and that the times in [11] and [13] were not given explicitly. In the case of [13], we derived them from the coil embolization example where authors state constituent times of a simulation of 40 Cosserat elements. Adding these and linearly extrapolating from 40 (2.26ms) to 100 elements gives an approximated time of 5.65ms. [11] only gives the times for length of rod without stating the underlying discretization. Thus, we made an assumption that the authors, like us, used 1 Cosserat element per 1mm of rod's length. We linearly extrapolated to 50, 100 and 1000 elements (e.g. 7.85ms / 900mm → 8.25ms / 1000 elements).

Table 2. Comparison of computational times with other models. *Approximated times.

Model	PC CPU	Time 50 El. (ms)	Time 100 El. (ms)	Time 1000 El. (ms)	Speed-up 50 El.	Speed-up 100 El.	Speed-up 1000 El.
Our inexten-sible mod	Core2 2.66 GHz	0.0868	0.1473	1.089	x1.00	x1.00	x1.00
Our original CoRdE impl.	Core2 2.66 GHz	0.1062	0.189	1.497	x1.22	x1.28	x1.37
Original CoRdE [12]	Xeon 3.80 GHz	0.069	0.131	1.24	x0.79	x0.89	x1.14
Wen Tang [11]	Core2 2.80 GHz	0.305*	0.61*	8.72*	x3.51*	x4.14*	x8.00*
Inext. CoRdE [13]	Core2 3.00 GHz	2.82*	5.65*	56.5*	x32.49*	x38.36*	x51.9*

Considering 1000 elements, our CoRdE implementation was slightly slower and our inextensible modification was slightly faster than the original CoRde [12]. However, there is more than 1GHz difference in CPU speed in favor of [12]. If our approximations were correct, our inextensible modification was x8.0 and x51.9 time faster than two other inextensible approaches ([11] and [13]). However, these models have some advantages over ours, e.g. the elimination of the penalty method in parallel constraints.

4   Conclusions

We present an instrument model for cardiovascular interventions. The behaviour of the virtual catheter and guidewire is based on a fast inextensible Cosserat rod implementation that allows for efficient modelling of bending, stretching and twisting phenomena, as well as guaranteeing almost immediate response to user manipulations, even for long instruments. The mechanical parameters of six guidewires and three catheters were optimised with respect to their real counterparts scanned in a silicone phantom using CT. The implementation allows the simulator to run efficiently on an off-the-shelf PC or laptop, significantly exceeding the minimum required haptic interactive rate.

Results show the parameter-optimised virtual instruments to exhibit near submillimetre accuracy, with errors likely to be caused by the accidental rotations and resulting torsion introduced during the insertion of real instruments into the silicone phantom. In contrast to the "perfect" simulated environment, such inaccuracies cannot be avoided in a real situation. The lack of friction between the virtual instruments and vessel walls is also likely to add to the overall error.

Future planned enhancements include improving the instrument's realism by adding Coulombian friction, which may affect instrument path when navigating inside the vessel, improving stability and further increasing the performance by developing a massively parallel GPU implementation. We are also developing a VR simulator supporting a complete angioplasty procedure, including instrument manipulation with force-feedback via a VSP haptic device (*www.mentice.com*), contrast medium propagation, balloon inflation, stent deployment and realistic x-ray visualization.

Acknowledgments. This work was supported in part by grants from the EPSRC, the UK Engineering and Physical Science Research Council, the RCUK Digital Economy Programme, the London Deanery STeLI initiative and Health Education North West London. The work from F. Martinez-Martinez has been funded by the Spanish Government under the FPI grant BES-2011-046495 and the grant EEBB-I-13-07588.

References

1. WHO, Global status report on noncommunicable diseases 2011: Geneva
2. Bridges, M., Diamond, D.L.: The financial impact of teaching surgical residents in the operating room. Am. J. Surg. 177(1), 28–32 (1999)
3. de Montbrun, S.L., Macrae, H.: Simulation in surgical education. Clin. Colon. Rectal Surg. 25(3), 156–165 (2012)
4. Gould, D.A., et al.: Simulation devices in interventional radiology: validation pending. J. Vasc. Interv. Radiol. 20(7 suppl.), S324–S325 (2009)
5. Wang, F., et al.: A computer-based real-time simulation of interventional radiology. In: Conf. Proc. IEEE Eng. Med. Biol. Soc., vol. 2007, pp. 1742–1745 (2007)
6. Luboz, V., et al.: Real-time guidwire simulation in complex vascular models. The Visual Computer 25(9), 827–834 (2009)
7. Duriez, C., et al.: New approaches to catheter navigation for interventional radiology simulation. Comput. Aided Surg. 11(6), 300–308 (2006)
8. Alderliesten, T., et al.: Modeling friction, intrinsic curvature, and rotation of guide wires for simulation of minimally invasive vascular interventions. IEEE TBME 54(1), 29–38 (2007)
9. Pai, D.K.: STRANDS: Interactive simulation of thin solids using cosserat models. Computer Graphics Forum 21(3), 347–352 (2002)
10. Antman, S.: Nonlinear Problems of Elasticity. Springer (1995)
11. Wen, T., et al.: A stable and real-time nonlinear elastic approach to simulating guidewire and catheter insertions based on Cosserat rod. IEEE Trans. Biomed. Eng. 59(8), 2211–2218 (2012)

12. Spillmann, J., Teschner, M.: CORDE: Cosserat Rod Elements for the Dynamic Simulation of One-Dimensional Elastic Objects. In: Symposium on Comp Animation 2007: ACM Siggraph/ Eurographics Symposium Proceedings, pp. 63–72 (2007)
13. Spillmann, J., Harders, M.: Inextensible elastic rods with torsional friction based on Lagrange multipliers. Computer Animation and Virtual Worlds 21(6), 561–572 (2010)
14. Duratti, L., et al.: A Real-Time Simulator for Interventional Radiology. In: VRST 2008 Proceedings of the 2008 ACM Symposium on Virtual Reality Software and Technology (2008)
15. Tang, W., et al.: A realistic elastic rod model for real-time simulation of minimally invasive vascular interventions. Visual Computer 26(9), 1157–1165 (2010)
16. Bergou, M., et al.: Discrete elastic rods. ACM Transactions on Graphics 27(3) (2008)
17. Spillmann, J.: CORDE: Cosserat Rod Elements for the Animation of Interacting Elastic Rods PhD Thesis (2008)
18. Chipperfield, A., et al.: Genetic Algorithm TOOLBOX For Use with MATLAB (1994)

3D Interactive Ultrasound Image Deformation for Realistic Prostate Biopsy Simulation

Sonia-Yuki Selmi[1], Emmanuel Promayon[1],
Johan Sarrazin[1,2], and Jocelyne Troccaz[1,*]

[1] UJF-Grenoble 1 / CNRS / TIMC-IMAG UMR 5525
Grenoble, F-38041, France
[2] KOELIS SAS, 5. av. du Grand Sablon, La Tronche, F-38700, France
FirstName.Name@imag.fr

Abstract. Realistic medical procedure simulators improve the learning curve of the clinicians if they can reproduce real conditions and use. This paper describes the improvement of a transrectal ultrasound guided prostate biopsy simulator by adding the simulation of real-time prostate movements and deformations. A discrete bio-mechanical model is used to modify a 3D texture of an ultrasound image volume in order to quickly simulate the actual displacements and deformations. This paper describes this model and presents how the mesh deformation is used to induce the UltraSound volume deformation. The validation of the method is based on both a quantitative and a qualitative assessment. Experimental images acquired on a phantom are compared using mutual information metrics to the resulting generated images. This comparison shows that the proposed method offers realistic deformed 3D ultrasound images at interactive time. The method was successfully integrated to improve the transrectal ultrasound simulator.

1 Introduction

Prostate cancer is the second most common cancer worldwide in men [1]. To confirm diagnosis, prostate biopsy procedures are performed to obtain and analyze tissue samples of the gland. Conventional biopsies are performed under TransRectal Ultrasound (TRUS) guidance. In clinical practice, a 12-core biopsy protocol is usually performed, using an end-fire imaging probe. To obtain a complete and accurate cancer diagnosis, these 12 samples have to be well-distributed and located in 12 different 3D anatomical zones of the prostate. The procedure is challenging because it requires a good understanding of ultrasound and an accurate spatial awareness of the prostate which is a small almost spherical organ (about 4 cm in diameter). The clinician has to perform a biopsy by positioning the samples only using information from the visualized UltraSound (US) images.

* This work was supported by French state funds managed by the ANR within the PROSBOT project and the Investissements d'Avenir programme (Labex CAMI) under reference ANR-11-LABX-0004 and by INSERM CHRT (grant J. Troccaz)

F. Bello and S. Cotin (Eds.): ISBMS 2014, LNCS 8789, pp. 122–130, 2014.

They have to understand the displayed US images in order to evaluate the actual position of the probe and perform the recommended biopsy.

We first designed a virtual reality simulator combining a learning environment (including a scoring system) and a clinical case database for image-guided prostate biopsy [2]. Its main purpose was to provide a real-time ultrasound simulator dedicated to prostate biopsy giving a score which evaluates how well a biopsy is distributed and accelerates the training phase. In this simulator, a laptop computer is connected to a haptic interface Phantom Omni (Sensable Devices Inc., MA, USA), used as a motion tracker. It is enhanced by a probe mock-up and a silicon-based structure representing the rectum, see Fig. 1a. The probe mock-up position and orientation is used to reslice a 3D recorded US volume. The simulator has a database of anonymized patient volumes and clinical data collected from a UroStation system (Koelis, France). A biopsy feedback is given by visualizing sample positions relative to the 3D image of the prostate and by computing a corresponding score. Initial validation proved its reliability, validated its face and content but it also showed that realism matters, especially for experts [3]. One of the main limitations highlighted by the experts was the absence of image deformation induced by the probe displacement. Because the prostate is a soft and highly-mobile organ, understanding deformation induced by the ultrasound probe manipulation is an important part of the learning process as the deformation will directly influence the sample positioning. The aim of the new simulator version is to include real-time realistic image deformation in order to have a complete prostate biopsy commercial simulator.

This paper presents a new version of the simulator that integrates interactive simulation of the tissue deformation.

2   Context

Ultrasound simulation is a commonly studied theme because on one hand, ultrasound imaging is a widely used image modality during surgical procedure and on the other hand, ultrasound image understanding is quite challenging, which justifies the need to develop simulators. Additionally, ultrasound manipulation strongly depends on the operator's experience. Ultrasound imaging is well adapted to soft tissue examination. Modeling ultrasound image deformation combines two challenges: to produce a realistic US image and to obtain realistic deformation of organs due to probe manipulation. Two main approaches try to address the first challenge: *a)* the generative approach, which consists in modeling the wave propagation, and *b)* the interpolation approach, which consists in acquiring dense 3D ultrasound volume from which arbitrary images can be generated. Two main approaches try to address the second challenge: the biomechanical continuous approach and the computational discrete one.

To our knowledge, few devices were designed for the simulation of prostate biopsy procedures. Chalasani et al. [4] reported the development and validation of a Virtual Reality TRUS guided prostate biopsy simulator, although without computing any deformation. While a lot of work focuses on only one of the two challenges [5] [6], combining US image generation with a biomechanical model is another challenge on its own. Goksel et al. [7] presents a brachytherapy training simulator with 2D simulated ultrasound and tissue deformation using the interpolating FEM method. D'Aulignac et al. [8] presents on a physical model for thrombosis diagnosis based on a mass-spring model and on 2D US image simulation using an interpolation approach. Vidal et al. [9] present a simulator of ultrasound guided needle puncture using virtual 2D multiplanar image reconstructed from input CT data.

3   Methods

We propose a method that combines a discrete biomechanical model combined with a 3D elastic texture in order to dynamically re-slice the patient's 3D prostate volume to obtain ultrasound image deformation, see Fig. 1b. The discrete model is based on shape memory [10]. The displacement of the probe is used as a displacement constraint by the model, which in turn computes the deformation on the physical object. Anonymized 3D US images acquired during actual biopsy of real patients are used as 3D textures. The biomechanical model deformation is used as a 3D texture mapping, which after interpolation and reslicing in the probe direction, generates the deformed US images. For each virtual biopsy sample in the simulator, the corresponding needle position is mapped in the undeformed biomechanical model in order to provide the actual 3D position of the performed biopsy in the undeformed gland. The deformation induced by the needle during the biopsy gun firing procedure is not taken into account in the model and neglected in the simulator.

During simulation, we ensure that stability is reached before performing the image generation step by controling the kinetic energy of the nodes.

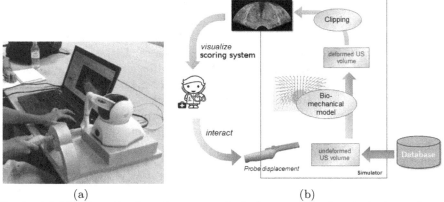

(a)   (b)

Fig. 1. (a) BiopSym simulator is composed of a laptop and a haptic device, enhanced by a structure simulating the rectum. (b) General architecture of the new simulator

Biomechanical Model. Our discrete soft tissue model has been already used in the context of prostate surgical interventions in [10]. It is described by a set of nodes, defined by their positions and their neighbors. In the model, elasticity is computed using shape memory. Each node is located using a shape memory function that depends on its position at rest shape $\mathbf{P}^0$ (undeformed mesh) and a generalized barycentric coordinates system based on its neighbor positions. During the simulation, the shape memory function directly computes the position of the node shape attractor $\mathbf{P}^*$. The shape memory force simulating the local elastic force is defined as $\mathbf{F_e}^t = -k_e\mathbf{P}^t\mathbf{P}^*$, where $\mathbf{P}^t$ is the node current position, and k_e is the elasticity parameter. The shape memory function is defined so that $\mathbf{P}^* = \mathbf{P}^0$ when there is no local deformation of the mesh relative to $\mathbf{P}$ and its neighbors.

In this application, the model geometry directly represents the 3D tissues imaged by the US probe, and therefore can be defined as a regular 3D grid, see Fig. 2 top left. The use of a regular grid yields to a limited number of shape memory configurations: a node is either a corner of the grid, on an edge, on the face or inside the grid. This limited number of configurations is one of the key points for obtaining an optimized elasticity computation time as the shape memory function has only four configurations. For instance, for a node inside the grid, the attractor $\mathbf{P}^*$ is always defined as the isobarycenter of the neighbors.

The displacement of all the nodes on the face opposite to the probe position is fixed. The US probe is simulated by a sphere half inserted in the grid of the same size as the actual end-fire US probe ($\varnothing = 13mm$), see Fig. 2 top left. During the simulation, the nodes that are initially enclosed in the sphere are moved with the probe displacement recorded by the Phantom Omni. Detected collisions with other nodes are processed using a radial projection constraint.

The following algorithm details the main step of the simulation and provides the nodal displacements at each iteration from the probe displacement:

for each iteration **do**
    get probe position and apply radial projection constraint
    for all nodes **do**
        compute attractor positions (4 possible configuration)
        compute forces
        compute internal forces
    end for
    for all nodes **do**
        integrate forces (e.g., using a explicit Verlet integrator)
    end for
end for

Image Deformation. The next step consists in applying the deformation to the 3D ultrasound volume using the nodal displacements, see Fig. 2. To deform the input volume, a non-linear warp transformation is defined on a 3D uniform grid and computed from the nodal displacements. In order to optimize the execution time, the warping grid is a subsample of the biomechanical grid. The deformed volume is then resliced along the image plane defined by the virtual probe position and orientation.

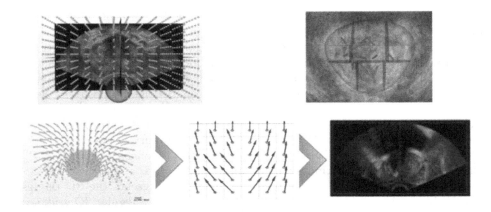

Fig. 2. Top left: US volume and nodes grid at initial step. Top right: Biopsy samples represented in the 3D prostate volume. Bottom: The algorithm starts by using a grid of physical nodes, from the displacements of those physical nodes a grid containing control points is created. This grid is then used to compute image voxels displacements.

Biopsy. In order to provide the 3D mapping of the complete biopsy and the associated score, the sample positions are computed from the positions taken in the deformed US volume and mapped to the undeformed volume. We used the mean value coordinates method [11] in order to map the sample position from the deformed volume to the undeformed biomechanical mesh.

Validation Process. Our objective is to obtain realistic deformed images. The validation process is both qualitative and quantitative. Qualitative assessment is achieved by comparing real procedure prostate deformation with the images generated by the simulator, based on real patient data. Fig. 3 compares the images obtained with and without the image deformation. The absence of quantified data related to US probe position during US volume acquisition on real patients led us to perform experiments on prostate phantom. From a quantitative point of view, the validation is based on image similarity comparisons between original and deformed images. This quantitative validation requires original and deformed images as well as each associated probe position. We used a set of 4 volumes obtained from a prostate tissue-mimicking deformable PVC phantom [12] with a size of $87 \times 100 \times 145mm$. Polaris localizer (NDI, Waterloo, Canada) markers are attached to the TRUS probe to track its position. The initial underformed volumes and final deformed volumes resulting from each probe displacements within the phantom are recorded alongside the corresponding probe final positions. The same probe positions are then simulated by interpolating them by steps of $1mm$ and the resulting simulated deformed images are saved. A region of interest centered on the prostate is defined in order to compare the simulated with the real deformed volume using mutual information [13]. The expected result is an increase of mutual information between the simulated volumes and the real deformed volume as the simulated probe position is nearing the final position step-by-step.

Implementation. The simulator, the image deformation methods and the validation process are implemented in C++ using the framework CamiTK[1] [14], ITK, VTK and OpenHaptics libraries. Mutual information is computed as explained in [15]. The size of the regular mesh for the biomechanical model is $20 \times 20 \times 20$ (8,000 nodes). The simulation time step and elasticity stiffness k_e were chosen to optimize stability. The size of the warping grid is $10 \times 10 \times 10$ (1,000 control points). The size of the images of the phantom are $105.6 \times 105.6 \times 79.1$ mm with a resolution of $199 \times 199 \times 199$ voxels and the probe displacements are in a range of $[20.9 - 27.3]mm$.

4   Results and Discussion

Without any deformation (only reslicing), the computing time is 70 ms on a Intel Core i7.3740QM 2.70 Ghz Pentium computer. Our method can generate slices in 178 ms on average. For each step, the relative computation time is 15% for the biomechanical computation, 45% for image deformation and 40% for the reslicing. One iteration of the verlet integrator on our discrete biomechanical model takes 0.6 ms. Thanks to the discrete model and the dynamic integrator, the computation time of one iteration is very low. Therefore the kinetic energy threshold is quickly reached when the probe movement is small (about 40 iterations), which provides a smooth user interaction.

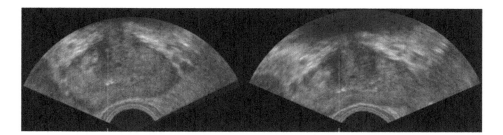

Fig. 3. The figure presents the 2D image of the undeformed volume (with no probe inducement) and the 2D image of the deformed volume (with a horizontal probe inducement of 13 mm). The white line represents the needle position during the biopsy fire.

Some of the acquired and simulated images are presented in Fig. 4a and b. The results can be visualized at http://youtu.be/wzjCk2fPwYE.

The purpose of this work and its first validation was to evaluate the feasibility of the method in terms of calculation time and increase the realistic visual quality. The discrete biomechanical model was not build to reproduce very accurate

---

[1] See http://camitk.imag.fr

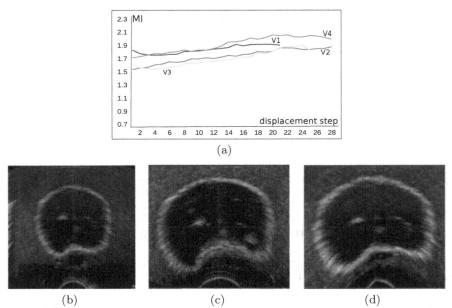

(a)

(b)   (c)   (d)

Fig. 4. (a) Computed mutual information on voxel values at each step for the 4 volumes (b) The initial input is an acquisition of the phantom without probe inducement (c) Slice of deformed volume acquired (d) Generated image

behaviour: k_e is not directly linked to physical properties and the model does not simulate the heterogeneous tissues nor their interaction. Despite this lack of physical realism, and as all constraints are based on displacements, we found that the equilibrium positions of each frame were not dependent of the elasticity parameter k_e. The main direction of improvement, especially in terms of predictive abilities on the biopsy score would be to model different organ structures, model the needle interaction with the tissue and to improve the probe interaction. Moreover, some boundary conditions can be easily added by applying a constraint of null displacement to any node, in order to simulate for instance the pelvic bone. The resulting generated images seem qualitatively satisfactory; the prostate has the characteristic bean-shape as seen during TRUS prostate examination, see Fig. 4. The main limitations of the qualitative experiments are *a)* the unknown intermediary images and probe displacements and *b)* the rather large probe displacements compared to a real procedure, where experts estimate a normal probe displacements to be around 10 mm. We plan to do new experiments on the phantom in order to record intermediate deformed volumes with smaller and more realistic probe displacements to improve the qualitative validation relevance. Quantitatively, the computed mutual information between each generated image and original final image for the set of 4 volumes shows that the image similarity increases, see Fig. 4b. This is a satisfactory result considering the fact that we did not use a classical finite element biomechanical model. While the biomechanical grid resolution has no strong effect on the time performance, the main bottleneck comes from the dimensions of the input volume and the warping grid resolution. The limitation due to the input volume directly comes from the choice of the interpolative approach. The wraping grid resolution could be optimized based on more experimental evaluation. Finally, it could be useful to integrate haptic force generation to take into account the model data in order to add realism.

5 Conclusion

In this paper, a fast ultrasound image deformation method based on a discrete biomechanical model that deforms real patient 3D volumes was presented. The implementation was successfully integrated into a learning simulator, to improve on classical simulators such as those described in [3,4]. A first validation was performed both qualitatively and quantitatively, showing an increase in the visual realism. The next step will be to evaluate the impact of our simulator, with and without including the image deformation, compared to traditional training.

References

1. Baade, P.D., Youlden, D.R., Krnjacki, L.J.: International epidemiology of prostate cancer: geographical distribution and secular trends. Mol. Nutr. Food Res. 53(2), 171–184 (2009)
2. Selmi, S.Y., Fiard, G., Promayon, E., Vadcard, L., Troccaz, J.: A virtual reality simulator combining a learning environment and clinical case database for image-guided prostate biopsy. In: 26th IEEE International Symposium on Computer-Based Medical Systems, Porto, Portugal (2013)
3. Fiard, G., Selmi, S., Promayon, E., Vadcard, L., Descotes, J.-L., Troccaz, J.: Initial validation of a virtual-reality learning environment for prostate biopsies: realism matters! Journal of Endourology 28(4), 453–458 (2014)
4. Chalasani, V., Cool, D.W., Sherebrin, S., Fenster, A., Chin, J., Izawa, J.I.: Development and validation of a virtual reality transrectal ultrasound guided prostatic biopsy simulator. Can Urol. Assoc. J. 5(1), 19–26 (2011)
5. Wein, W., Khamene, A., Clevert, D.-A., Kutter, O., Navab, N.: Simulation and Fully Automatic Multimodal Registration of Medical Ultrasound. In: Ayache, N., Ourselin, S., Maeder, A. (eds.) MICCAI 2007, Part I. LNCS, vol. 4791, pp. 136–143. Springer, Heidelberg (2007)
6. Jahya, A., Herink, M., Misra, S.: A framework for predicting three-dimensional prostate deformation in real time. The International Journal Of Medical Robotics And Computer Assisted Surgery 9, 52–60 (2013)
7. Goksel, O., Sapchuk, K., Morris, W.J., Salcudean, S.E.: Prostate Brachytherapy Training with Simulated Ultrasound and Fluoroscopy Images. IEEE Trans. Biomedical Engineering 60(4), 1002–1012 (2013)
8. d'Aulignac, D., Laugier, C., Troccaz, J., Vieira, S.: Towards a realistic echographic simulator. Medical Image Analysis 10(1), 71–81 (2006)
9. Vidal, F.P., John, N.W., Healey, A.E., Gould, D.A.: Simulation of ultrasound guided needle puncture using patient specific data with 3D textures and volume haptics. Comp. Anim. Virtual Worlds 19, 111–127 (2008)

10. Marchal, M., Promayon, E., Troccaz, J.: Simulating prostate surgical procedures with a discrete soft tissue model. In: Navazo, I., Mendoza, C. (eds.) Eurographics Workshop in Virtual Reality Interactions and Physical Simulations, I, pp. 109–118 (2006)

11. Floater, M.S., Kos, G., Reimers, M.: Mean value coordinates in 3D. Comput. Aided Geom. Design 22(7), 623–631 (2005)

12. Hungr, N., Long, J.A., Beix, V., Troccaz, J.: A realistic deformable prostate phantom for multimodal imaging and needle-insertion procedures. Med. Phys. 39(4), 2031–2041 (2012)

13. Maes, F., Collignon, A., Vandermeulen, D., Marchal, G., Suetens, P.: Multimodality image registration by maximization of mutual information. IEEE Transactions on Medical Imaging 16, 187–198 (1997)

14. Fouard, C., Deram, A., Keraval, Y., Promayon, E.: CamiTK: A Modular Framework Integrating Visualization, Image Processing and Biomechanical Modeling. In: Payan, Y. (ed.) Soft Tissue Biomechanical Modeling for Computer Assisted Surgery, pp. 323–354. Springer, Heidelberg (2012)

15. Mattes, D., Haynor, D.R., Vesselle, H., Lewellen, T.K., Eubank, W.: Nonrigid multimodality image registration. In: Sonka, M., Hanson, K.M. (eds.) Medical Imaging: Image Processing. Proc. SPIE, vol. 4322, pp. 1609–1620. SPIE Press, Bellingham (2001)

Interactive Deformation
of Heterogeneous Volume Data

Rosell Torres[1], Jose M. Espadero[1], Felipe A. Calvo[2],
and Miguel A. Otaduy[1]

[1] URJC Madrid, Móstoles (Madrid), Spain
[2] HGU Gregorio Marañón, Madrid, Spain

Abstract. This paper presents a method to interactively deform volume images with heterogeneous structural content, using coarse tetrahedral meshes. It rests on two major components: a massively parallel algorithm for the rasterization of tetrahedral meshes, and a method to define a coarse deformable tetrahedral mesh from the homogenization of a fine heterogeneous mesh. We show the potential of the method for training and planning applications through two examples: an abdominal CT exploration and the alignment of breast CT and MRIs.

1 Introduction and Related Work

Training and planning of surgical interventions on soft anatomical parts requires tools to deform 3D models of such anatomical parts. Deformable models have also become an integral part of advanced registration [10] or segmentation algorithms [12]. The traditional pipeline to produce deformable anatomical models is to acquire volumetric images of the anatomy, segment the desired anatomical parts, mesh them, and apply some deformation method such as FEM. This pipeline suffers several problems, most notably: lack of robustness of automatic segmentation, high computational cost of FEM to capture heterogeneous anatomy, and loss of the anatomical detail present in the original volume image.

In this paper, we present a method for the interactive deformation of heterogeneous anatomy built directly on the input volume images. Our method stands on two main components.

Homogenization: Given a fine discretization of a heterogeneous medium, homogenization is the process of determining parameters of a coarse discretization that best matches the behavior of the fine discretization. Kharevych et al. [6] proposed an approach that probes the fine and coarse meshes with linear forces, and then computes coarse elastic parameters such that the coarse mesh preserves the elastic energy of the fine mesh. Nesme et al. [9] proposed an approach that sets arbitrary kinematic constraints, and computes coarse elastic properties such that the displacements of coarse nodes match under both coarse and fine discretizations. Although not exactly a homogenization method, Faure et al. [2] introduced a discretization technique based on a sparse set of frames, where the deformation of the frames is transmitted to the object's volume using nonlinear shape functions that account for material heterogeneity.

F. Bello and S. Cotin (Eds.): ISBMS 2014, LNCS 8789, pp. 131–140, 2014.

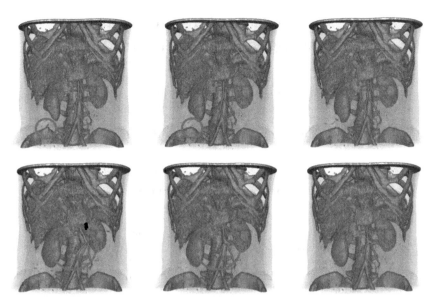

Fig. 1. From left to right: Coarse mesh with standard FEM solver; coarse mesh with our method; and fine mesh with standard solver. Top row: In the standard coarse mesh, deformations on the kidney are transferred to the hip due to insufficient mesh resolution. Bottom row: In the standard coarse mesh, the aorta appears softer due to the surrounding air. Our method, instead, enjoys subelement accuracy and retains the heterogeneity of the fine mesh.

In Section 2, we present a homogenization method that supports corotational elasticity and simplifies the construction of the coarse FEM problem in contrast to previous methods. As shown in Fig. 1 and Fig. 2, our method enables fast deformation using coarse tetrahedral meshes but capturing subelement heterogeneity. In Section 3, we describe the homogenization of two types of external forces: gravity and user interaction.

Volume Rasterization: One outstanding feature of our method is the possibility to deform volume data interactively. As an alternative, Nesme et al. [8] deformed planar slices with semi-transparent textures, and rendered them front to back. More similar to our work, Goksel and Salcudean [4] mapped the deformation of a tetrahedral mesh to the volume using texture mapping, but they followed a slow scanline approach, which was interactive only for 2D images.

We adapt the recent tetrahedral rasterization algorithm of Gascon et al. [3], a massively parallel algorithm to interactively rasterize tetrahedral meshes, using the original volume image as a 3D texture map. We define the deformation of the volume image in two steps: first from coarse tetrahedra to fine tetrahedra according to nonlinear shape functions, and then from fine tetrahedra to image voxels according to barycentric interpolation. Our approach deforms the full volume data, which allows online modification of volume rendering settings, as well as runtime volume ray casting for isosurface rendering.

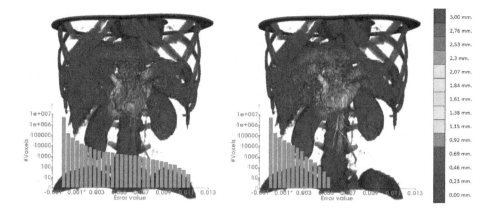

Fig. 2. Illustrations of per-voxel error and histograms of error distribution, comparing a standard coarse FEM simulation (left) with our homogenization approach (right). Error is measured w.r.t. a fine FEM simulation with double resolution. Please see the results section for more details.

Results: We show the potential of our method on two preliminary applications: the exploration of an abdominal CT, and the alignment of breast CT and MRIs. In both examples, we demonstrate interactive deformations of the volume images while capturing the heterogeneity of the underlying anatomy.

2   Deformation Model

This section describes our method for interactive deformation of volume images. The method uses two tetrahedral meshes, one of fine resolution and the other of coarse resolution. The deformation of the underlying anatomy is computed by solving an FEM elasticity problem on the coarse mesh. The fine mesh acts as an intermediate representation between the coarse mesh and the volume image at two stages. During preprocessing, it serves to define the FEM problem on the coarse mesh following a homogenization approach. At runtime, it serves to transmit the nonlinear deformations of the coarse mesh to the volume image.

2.1   High-Resolution Elasticity

Given a fine tetrahedral mesh that surrounds the volume image, we compute nodal masses and a fine stiffness matrix from image intensity values. Note that we never solve the actual fine elasticity problem; this is formulated only as an intermediate step for homogenization. To formulate the stiffness matrix, we adopt a linear isotropic elasticity formulation with linear basis functions. Thanks to the use of linear elasticity, the base stiffness matrix of fine tetrahedra can be precomputed, and the use of linear basis functions (i.e., barycentric interpolation) is convenient for the rasterization algorithm defined later.

Given a volume image, we first assign mass-density and elasticity properties (Young modulus and Poisson ratio) at voxel resolution. In our examples, we inspected the volume using a narrow transfer function, and associated voxel intensity ranges of major anatomical elements to default parameters.

To define nodal masses, we integrate mass-density over the voxels included in each tetrahedron, and we distribute one fourth of the tetrahedron mass to each of its four nodes. To compute the stiffness matrix of each tetrahedron, we simply adhere to basic principles and integrate the material tensor over the tetrahedron volume. With constant material parameters per voxel, this integral translates into a sum over all voxels in each tetrahedron, which yields the element stiffness $\mathbf{K}_e = v \, \mathbf{B}_e^T (\sum_i \mathbf{E}_i) \mathbf{B}_e$, where v is the voxel volume, $\mathbf{B}_e$ is a matrix of derivatives of shape functions, and $\mathbf{E}_i$ is each voxel's material tensor, formed from the voxel's Young modulus and Poisson ratio.

2.2   Homogenized Corotational Elasticity

Given a linear elastic problem on the fine mesh, we pose homogenization as the definition of an equivalent linear elastic problem on the coarse mesh. Homogenization requires the definition of: (i) (homogenized) stiffness matrices for coarse tetrahedra that best describe the fine heterogeneity, and (ii) nonlinear shape functions to compute the displacements of fine nodes from coarse nodes.

We describe first the computation of the homogenized stiffness matrix for each coarse tetrahedron. As shown in Fig. 3, for each coarse tetrahedron we define its corresponding fine submesh using all fine tetrahedra that intersect with the coarse tetrahedron. To ease the homogenization procedure, we set the following constraint on the coarse and fine meshes: coarse nodes must constitute a subset of the fine nodes. Then, we define as $\mathbf{u}_c$ and $\mathbf{u}_f$ (resp. $\mathbf{F}_c$ and $\mathbf{F}_f$) the displacements (resp. forces) of the coarse nodes and the rest of the fine nodes in the submesh. The fine linear elastic problem for each coarse tetrahedron can be formulated as:

$$\begin{pmatrix} \mathbf{K}_{cc} & \mathbf{K}_{cf} \\ \mathbf{K}_{fc} & \mathbf{K}_{ff} \end{pmatrix} \begin{pmatrix} \mathbf{u}_c \\ \mathbf{u}_f \end{pmatrix} = \begin{pmatrix} \mathbf{F}_c \\ \mathbf{F}_f \end{pmatrix}. \tag{1}$$

Assuming that forces are applied only on coarse nodes, i.e., $\mathbf{F}_f = 0$, the displacements of coarse nodes are:

$$\mathbf{u}_c = \left(\mathbf{K}_{cc} - \mathbf{K}_{cf} \, \mathbf{K}_{ff}^{-1} \, \mathbf{K}_{fc} \right)^{-1} \mathbf{F}_c. \tag{2}$$

From this expression, we directly derive the homogenized stiffness matrix that relates coarse forces to coarse displacements:

$$\mathbf{K} = \mathbf{K}_{cc} - \mathbf{K}_{cf} \, \mathbf{K}_{ff}^{-1} \, \mathbf{K}_{fc}. \tag{3}$$

Intuitively, with this homogenized stiffness matrix, the displacements of coarse nodes are matched when forces are applied only on coarse nodes. And it turns out that this stiffness matrix is nothing else but the Schur complement of the fine submatrix, and it also matches the result of *condensation* [1], although applied

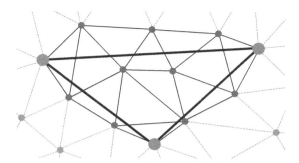

Fig. 3. 2D example showing the submeshes involved in the homogenization process. For each coarse triangle (tetrahedron in 3D), shown in thick line, we define a fine submesh using all fine triangles, shown in thin line, that intersect the coarse triangle. In green we show the coarse nodes, and in red the rest of the fine nodes in this submesh.

to a different problem. Our result is similar to the one by Nesme et al. [9] with displacement constraints on coarse nodes, but our formulation is notably simpler and more computationally efficient. Their approach builds the homogenized stiffness in a hierarchical fashion, allowing them to preserve the topology of the fine mesh. We miss this feature, but instead avoid the need for multiresolution meshes, we accommodate partial segmentations, and avoid artifacts under large rotations thanks to the use of corotational elasticity.

Homogenized stiffness matrices of coarse tetrahedra are computed as a preprocess. At runtime, we solve a quasi-static deformation problem on the coarse mesh, which requires the assembly of the full coarse stiffness matrix. We use a linear corotational elasticity approach [7], where the stiffness matrices of individual elements are warped based on per-element rotations and then assembled together. With our method, deformations are fast and stable because the coarse matrix remains positive semi-definite (PSD) provided a high-res PSD matrix. Since we compute no accelerations, we do not need to homogenize the mass matrix.

2.3   High-Resolution Deformation

We compute the displacements of fine nodes in a per-element corotational setting using the nonlinear basis functions resulting from homogenization, and then average the contributions of different coarse tetrahedra. Thanks to the corotational formulation, our approach preserves rigid transformations.

Fig. 4 illustrates the computation of high-res deformations using a corotational formulation. Given the best-fit rotation $\mathbf{R}$ of a coarse tetrahedron, and the deformed and undeformed positions of its centroid, $\mathbf{c}$ and $\bar{\mathbf{c}}$ respectively, the displacement of each coarse node in the undeformed reference frame can be computed as:

$$\mathbf{u}_{c,i} = \mathbf{R}^T \left(\mathbf{x}_{c,i} - \mathbf{c} \right) - \left(\bar{\mathbf{x}}_{c,i} - \bar{\mathbf{c}} \right), \tag{4}$$

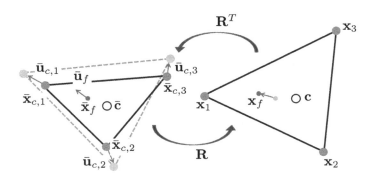

Fig. 4. 2D schematic illustration of the corotational computation of high-res deformations. On the left, a coarse undeformed triangle (tetrahedron in 3D), and on the right, the same triangle deformed. The deformed triangle is transformed to the undeformed setting based on a best-fit rotation $\mathbf{R}$ and the translation of the centroid $\mathbf{c}$. The displacements $\mathbf{u}_{c,i}$ of coarse nodes are computed in the undeformed setting (in dotted lines), and weighted using the nonlinear homogenized shape functions to obtain the displacement $\mathbf{u}_f$ of each fine node. This displacement is finally transformed to the deformed setting to produce the deformed position $\mathbf{x}_f$ of the fine node.

with $\mathbf{x}_{c,i}$ and $\bar{\mathbf{x}}_{c,i}$ the deformed and undeformed positions of the node.

From the definition of the fine elasticity problem (1), the displacements of fine nodes in the submesh of a coarse tetrahedron can be computed from the displacements of coarse nodes as

$$\mathbf{u}_f = -\mathbf{K}_{ff}^{-1}\mathbf{K}_{fc}\,\mathbf{u}_c. \tag{5}$$

From this expression, we can derive the definition of homogenized shape functions that relate fine to coarse displacements:

$$\mathbf{N} = -\mathbf{K}_{ff}^{-1}\mathbf{K}_{fc}. \tag{6}$$

Note that these shape functions are nonlinear in space.

Once the displacement of each fine node is obtained in the undeformed reference frame, it is simply transformed to the deformed setting, using again the corotational formulation:

$$\mathbf{x}_{f,i} = \mathbf{R}\left(\mathbf{u}_{f,i} + (\bar{\mathbf{x}}_{f,i} - \bar{\mathbf{c}})\right) + \mathbf{c}. \tag{7}$$

For a fine node that belongs to submeshes of multiple coarse tetrahedra, we weight the corresponding deformed positions using inverse-distance weighting based on distances to the centroids of those tetrahedra.

2.4   Volume Rasterization

To deform the volume image, and given the positions of the fine nodes, we apply the GPU-based volume rasterization algorithm by Gascon et al [3]. This algorithm rasterizes the original image into the target grid assuming a barycentric mapping inside each fine tetrahedron. To maximize efficiency, each tetrahedron is decomposed into cells, which are first culled on the CPU, and the cells that pass the culling test are rasterized in a massively parallel manner on the GPU.

3   External Forces

Interaction with the homogenized model can be obtained through the action of external forces. To preserve the behavior of the heterogeneous volume, these external forces should be defined on the fine mesh and then transmitted to the coarse mesh. Here we describe the homogenization of two types of external forces: gravity forces and pulling forces at arbitrary points.

3.1   Gravity Forces

Our method for computing gravity forces on the coarse mesh preserves the gravitational energy of the fine mesh, which is different from the homogenization of mass. With m_i the mass of each fine node, $\mathbf{u}_{c,i}$ the displacement, and $\bar{\mathbf{x}}_{c,i}$ the rest position, the gravitational energy is

$$V = -\sum_i m_i\, \mathbf{g}^T\left(\mathbf{u}_{c,i} + \bar{\mathbf{x}}_{c,i}\right). \tag{8}$$

The gravity force on a coarse node is defined as the negative gradient of the gravitational energy:

$$\mathbf{F}_j = -\frac{\partial V}{\partial \mathbf{u}_{c,j}}^T = \sum_i m_i\, \frac{\partial \mathbf{u}_{f,i}}{\partial \mathbf{u}_{c,j}}^T \mathbf{g} = \sum_i m_i\, \mathbf{N}_{i,j}^T\, \mathbf{g} = \mathbf{M}_j\, \mathbf{g}. \tag{9}$$

From the expression above, we observe that we can precompute a gravity-mass-matrix per coarse node as the mass-weighted sum of shape function weights:

$$\mathbf{M}_j = \sum_i m_i\, \mathbf{N}_{i,j}^T. \tag{10}$$

3.2   Isosurface Pulling

Since our deformation method deforms the full volume image interactively, it is possible to select arbitrary points of the deformed volume at any time. One interesting possibility is to pick a point on the isosurface of the visible volume through ray casting, and apply a pulling force at that point.

The displacement $\mathbf{u}_p$ of an arbitrary point $\mathbf{p}$ on the volume is defined based on barycentric interpolation of fine nodes, which are in turn interpolated from coarse nodes using the nonlinear shape functions. The displacement can be expressed as $\mathbf{u}_p = \beta\, \mathbf{N}\, \mathbf{u}_c$, with β a matrix of barycentric weights. Then, given a force $\mathbf{F}_p$ at point $\mathbf{p}$, this force is distributed to the coarse nodes using the same weights as for displacement interpolation, i.e.,

$$\mathbf{F}_c = \mathbf{N}^T \beta^T \mathbf{F}_p. \tag{11}$$

If multiple forces are applied at different points, they are added together.

4   Experiments and Results

All our examples were executed on a 3.1 GHz Quad-core Intel Core i7-3770S CPU with 16GB of memory, and a NVIDIA GTX670 graphics card. We used VTK [11] for interactive volume rendering and isosurface ray casting. We have tested our method in action on the following two examples.

4.1   Exploration of the Abdominal Cavity

Isosurface Pulling: The isosurface pulling functionality described in Section 3.2 enables planning interventions where anatomical parts must be moved aside to provide access to the region of interest. Examples include tumor resection in the retroperitoneum behind the kidney or in the adrenal gland hidden by the pancreas. We have evaluated several deformation examples produced by pulling directly from the volume data, in a highly intuitive manner.

Method Comparison and Performance: We have compared our deformation method with standard corotational FEM on the fine mesh and standard corotational FEM on the coarse mesh. With our method, a volume with 5M voxels ($256 \times 160 \times 122$), a fine mesh with $13,720$ tets, and a coarse mesh with $1,715$ tets, the abdomen is deformed at a rate of 86 ms/frame. FEM deformation takes 26% of the time, culling 32%, and rasterization 42%. Our method introduces a small performance penalty over the standard coarse deformation, which takes 44 ms/frame. The gain over the standard fine deformation is large, as this one takes 373 ms/frame. Fig. 1 shows artifacts suffered by the standard FEM model on the coarse mesh. For a maximum pulling deformation of 3.07 cm, the maximum error with the standard coarse simulation compared to a standard fine simulation is of 1.24 cm (40%). With our method instead, the maximum error is of 5.7 mm (18.5%). Fig. 2 compares visualizations of the error on the visible areas, as well as error histograms for a particular simulation frame.

4.2   Breast CT and MRI Alignment

Image-aided planning of breast cancer interventions is faced with the challenge of acquiring high-quality images in surgical supine position. One potential approach is to acquire an MRI in prone position, which is the preferred system for tumor identification and robust under respiration; acquire a CT in supine position, which is fast; and perform deformable registration of prone and supine acquisitions. As shown in Fig. 5, we have tested the use of our method to deform a prone MRI into the configuration of a supine CT. For this example, we assumed that the anatomy was in its rest configuration in the supine position, and we simply reversed the direction of gravity while rotating the model. Our method could be applied in an iterative algorithm combining registration and material estimation [5], with the power to execute extremely fast deformation of the whole 3D image data.

Adaptive Meshing: If a partial segmentation is available, it is possible to improve the quality of homogenization by designing an adaptive fine mesh, with no

Fig. 5. From left to right: Breast CT acquired in supine position; breast CT deformed to prone position under gravity using our method; breast MRI acquired in prone position

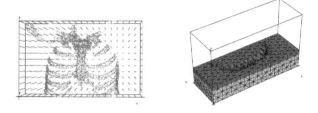

Fig. 6. Adaptive fine meshing. Left: Regularly sampled nodes together with nodes on the boundary of segmented bones. Right: Cut of the resulting tetrahedral mesh.

increase on the deformation cost of the coarse mesh. Fig. 6 shows an adaptive fine mesh for the breast model, combining regularly sampled nodes with nodes sampled on the surface of segmented bones. Fixing bone nodes becomes trivial with our homogenization approach. They are simply removed as degrees of freedom from the fine elasticity problem (1), and their shape functions are trivially set to zero.

Performance: With our method, a volume with 13.2M voxels ($524 \times 176 \times 144$), a fine mesh with $44,721$ tets, and a coarse mesh with $2,160$ tets, the breast model is deformed at a rate of 271 ms/frame.

5 Discussion and Future Work

We have introduced a method to interactively deform volume images. Interactivity is possible thanks to the use of coarse tetrahedral meshes for FEM deformation, and highly efficient rasterization of tetrahedra. Despite using coarse meshes, our homogenization approach captures subelement material heterogeneity, achieving accuracy close to that of fine meshes.

Our current homogenization method could be further improved in several ways. One is the extension to nonlinear elasticity, which would produce deformations with higher accuracy. Another one is handling topology changes.

We have demonstrated the potential for several clinical applications, but it would be interesting to explore this potential further, e.g., for planning of intraoperative radiotherapy in breast cancer surgery, exploiting the alignment of CT and MRIs. Other generic applications worth testing include deformable registration and segmentation.

Acknowledgements. We would like to thank the anonymous reviewers, Jorge Gascón for his help with the volume rasterizer, Carlos Illana for many suggestions, and María Jesús Ledesma, Javier Pascau and Laura Fernández for their help with the breast dataset. The human torso dataset was obtained from the OsiriX DICOM Viewers site. This work was supported in part by the Spanish Ministry of Economy (grants IPT- 2012-0401-300000 and TIN2012-35840), the EU FEDER fund, and the European Research Council (ERC-2011-StG-280135 Animetrics).

References

1. Bro-Nielsen, M., Cotin, S.: Real-time volumetric deformable models for surgery simulation using finite elements and condensation. Computer Graphics Forum 15(3), 57–66 (1996)
2. Faure, F., Gilles, B., Bousquet, G., Pai, D.K.: Sparse Meshless Models of Complex Deformable Solids. ACM Transactions on Graphics 30(4), Article No. 73 (2011)
3. Gascon, J., Espadero, J.M., Perez, A.G., Torres, R., Otaduy, M.A.: Fast deformation of volume data using tetrahedral mesh rasterization. In: Proc. of the ACM SIGGRAPH/Eurographics Symposium on Computer Animation, pp. 181–185 (2013)
4. Goksel, O., Salcudean, S.E.: B-mode ultrasound image simulation in deformable 3-d medium. IEEE Transactions on Medical Imaging 28(11), 1657–1669 (2009)
5. Han, L., Hipwell, J., Eiben, B., Barratt, D., Modat, M., Ourselin, S., Hawkes, D.: A nonlinear biomechanical model based registration method for aligning prone and supine mr breast images. IEEE Trans. on Medical Imaging 33(3), 682–694 (2014)
6. Kharevych, L., Mullen, P., Owhadi, H., Desbrun, M.: Numerical coarsening of inhomogeneous elastic materials. ACM Trans. on Graphics 28(3), 51:1–51:8 (2009)
7. Müller, M., Gross, M.: Interactive virtual materials. In: Proc. of Graphics Interface (2004)
8. Nesme, M., Faure, F., Payan, Y.: Accurate interactive animation of deformable models at arbitrary resolution. Intl. Journal of Image and Graphics 10(2) (2010)
9. Nesme, M., Kry, P.G., Jerábková, L., Faure, F.: Preserving topology and elasticity for embedded deformable models. ACM Trans. on Graphics 28(3), 52:1–52:9 (2009)
10. Parisot, S., Duffau, H., Chemouny, S., Paragios, N.: Joint tumor segmentation and dense deformable registration of brain MR images. In: Ayache, N., Delingette, H., Golland, P., Mori, K. (eds.) MICCAI 2012, Part II. LNCS, vol. 7511, pp. 651–658. Springer, Heidelberg (2012)
11. Schroeder, W., Martin, K., Lorensen, B.: The Visualization Toolkit: An object-oriented approach to 3D graphics, 3rd edn. Tech. rep., Kitware Inc. (2004), http://www.vtk.org
12. Uzunbaş, M.G., Chen, C., Zhang, S., Pohl, K.M., Li, K., Metaxas, D.: Collaborative multi organ segmentation by integrating deformable and graphical models. In: Mori, K., Sakuma, I., Sato, Y., Barillot, C., Navab, N. (eds.) MICCAI 2013, Part II. LNCS, vol. 8150, pp. 157–164. Springer, Heidelberg (2013)

Brain Ventricular Morphology Analysis Using a Set of Ventricular-Specific Feature Descriptors

Jaeil Kim[1,*], Hojin Ryoo[1,*], Maria del C. Valdés Hernández[2],
Natalie A. Royle[2], and Jinah Park[1,**]

[1] Department of Computer Science,
Korea Advanced Institute of Science and Technology,
291 Daehak-ro, Yuseong-gu, Daejeon, Republic of Korea
{threeyears,ryoo.hojin,jinahpark}@kaist.ac.kr
[2] Centre for Clinical Brain Sciences, University of Edinburgh,
Chancellor's Building, 49 Little France Crescent, Edinburgh EH16 4SB, UK
{M.Valdes-Hernan,nat.royle}@ed.ac.uk

Abstract. Morphological changes of the brain lateral ventricles are known to be a marker of brain atrophy. Anatomically, each lateral ventricle has three horns, which extend into the different parts (i.e. frontal, occipital and temporal lobes) of the brain; their deformations can be associated with morphological alterations of the surrounding structures and they are revealed as complex patterns of their shape variations across subjects. In this paper, we propose a novel approach for the ventricular morphometry using structural feature descriptors, defined on the 3D shape model of the lateral ventricles, to characterize its shape, namely width, length and bending of individual horns and relative orientations between horns. We also demonstrate the descriptive ability of our feature-based morphometry through statistical analyses on a clinical dataset from a study of aging.

1 Introduction

The brain lateral ventricles are cavities filled by cerebrospinal fluid (CSF) limited by an epithelial membrane called ependymal. The morphological deformations on the lateral ventricles have been related to neurodegenerative diseases [7] and cognitive decline [2]. It has also been reported that the ventricular volume and width may be predictors to determine the quality of brain surgical treatments, such as deep brain stimulation surgery [4] and ventricular catheter placement [20].

Deformations of the lateral ventricles have been investigated through various approaches that involve assessing the volume [3], ventricular surface [13] and medial thickness [10]. However, the analysis of the ventricular deformations is still

---

* Both authors contributed equally to this work.
** Corresponding author.

F. Bello and S. Cotin (Eds.): ISBMS 2014, LNCS 8789, pp. 141–149, 2014.

challenging. Since the lateral ventricles are essentially a brain cavity with irregular shape, its shape changes can be atypical and revealed as complex patterns across subjects. This not only applies to each horn individually, but also to the "C-shape" that curves from the temporal horn into the beginning of the frontal horn [18], which also varies across subjects. The atrium of the lateral ventricles is the place where the three horns meet, and it can also change depending on each horn's enlargement and on changes in the shape of the surrounding structures. Due to the adaptive capacity of the ventricular system filled by CSF, its structural changes may be related to the alteration of the surrounding parenchymal structures [13]. Therefore decomposing the structural changes of the lateral ventricle anatomically can lead to a better understanding of the ventricular deformations in relation to the surrounding structures.

For the morphology analysis of the brain ventricles, several indices quantifying the structural characteristics of the ventricles have been proposed. With respect to the interior and exterior of the lateral ventricles, these shape indices of the ventricles can reveal the atrophy of the interior structures of the medial temporal lobe or the global and lobar atrophy of the brain [8]. For instance, the radial width of the temporal horn was introduced to measure the enlargement of the temporal horn [8] and to investigate the atrophy of the hippocampus where the memory consolidates [5]. Some clinical studies reported that the atrophy of the frontal lobe is related with aging and cognitive decline [2,5], diabetes and drug addiction [13,15]. To trace the enlargement of the frontal horn of the lateral ventricle, the structural measures, such as ventricular angle and frontal horn radius, has been used [4,17] as well as 3D shape models of the ventricle. These structural indices can provide a straightforward description of the ventricular shape and its changes. However, the manual measurement of them makes it susceptible to human errors and imaging protocols.

By providing an automated modeling strategy that consistently characterizes the shape of the lateral ventricle across subjects, we are addressing the needs on computational analyses: intuitive shape description and reproducibility. Based on the surface mesh of the lateral ventricle, we propose a set of the explicit feature descriptors as a pre-defined layout of its shape description to quantify the structural characteristics (e.g. horn length, radius and bending) independently. We propose two types of feature descriptors: (1) individual features for quantifying the shape and size of each horn, and (2) inter-horn features for characterizing the C-shape and the atrium of the lateral ventricles. We also demonstrate the descriptive ability of our feature-based morphometry through the statistical analysis on a dataset from a study of aging.

2   Ventricular Shape Representation

We model the shape of lateral ventricles as a smooth surface mesh and its central skeleton. These model components form the basis of our approach allowing quantifying the structural characteristics of the individual horns and between the horns.

First of all, the surface mesh of lateral ventricle is constructed from binary masks, obtained from brain MR images, via a template-based surface modeling approach [14]. We construct a template mesh of the lateral ventricle using an average shape image, generated from the binary masks via a image registration based approach [12]. We optimally align the template mesh to each binary mask and propagate it to the image boundary while minimizing the distortion of its point distribution against arbitrary size variations. This method builds a pairwise correspondence between the template of average shape and the targets of each subject.

On the reconstructed meshes, we easily determine the tips of the individual horns, owing to the point correspondence with the template. They are used for the construction of the skeleton. For the skeleton construction, we use a Voronoi diagram-based approach where the centers of the maximal inscribed spheres (MIS) are defined within and along the surface mesh [1]. We used the vascular modeling toolkit (VMTK, Ver. 1.2) for this process. Technically, from two terminal points and the surface model, the skeleton is constructed by finding a path between two points, which minimizes the integral of the inverse radius of the MIS on the Voronoi diagram of the surface model [19].

In the ventricular shape representation, the most important structure is the central skeleton which consists of the atrium center and the horn skeleton (See Figure 1). The *atrium center* is defined as a center of the ventricular atrium region. To determine the atrium center, we first compute three tip-to-tip skeletons; the frontal-temporal, the temporal-occipital, and the occipital-frontal skeleton. Then, the atrium center is computed by finding an optimal position which has equal minimum distance to the three tip-to-tip skeletons on the Voronoi diagram. The atrium center is assigned to be a common terminal point for constructing the skeletons of the three horns, which are referred each as "(frontal/temporal/occipital) horn skeleton". The horn skeleton is the skeleton connecting the atrium center and the tips of each horn.

Other feature landmarks necessary to determine the domain of the feature measurement characterizing each horn's morphology are the *starting and ending points* for each horn-skeleton. The starting point is determined by the center of the MIS that passes through the atrium center in the horn skeleton, and the ending point is the center of the last MIS passing through the tip of each horn. We also define two vectors for each horn – *starting and ending vectors* – from the atrium center to the starting and ending points, respectively.

3   Structural Feature Descriptors for the Lateral Ventricle

On the representation model of the lateral ventricle, we define the two types of feature descriptors, the individual horn descriptors and the inter-horn descriptors, which quantify the geometric characteristics for the lateral ventricle.

As individual horn structural descriptors, we define the width, length, and bending of each horn with respect to its skeleton. The *width* descriptor informs how large the cross-section of the horn is. Once we determine the closest distance from each vertex of the surface mesh to the skeleton, we calculate the width at

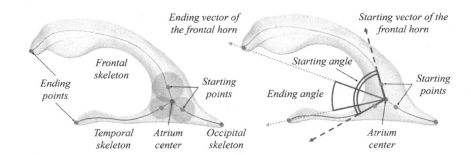

Fig. 1. Ventricular shape model with its skeleton and landmarks. The green and blue points indicate respectively the starting and ending point of each horn skeleton, and the red point indicates the atrium center. The starting vectors (dotted arrow) and the ending vectors (dashed arrow) are referred to measure the angles between two horns.

each sampled point of the skeleton as the average of the closest distance to the vertices of the surface mesh, which are nearest to the sampled points. The skeleton is segmented at a regular interval of a unit distance to determine the sampled points. We use the average value of the width at each sampled point and the width at the starting point as ventricular feature descriptors. The *length* of each horn refers to the geodesic length of the skeleton, which is calculated by summing the Euclidean distances between the sampled points of the skeleton. The *bending* of each horn is estimated as the ratio of the geodesic length to the Euclidean distance between the starting and ending points for each horn.

As inter-horn structural descriptors, we define the width at the atrium center and the angles representing the relative orientations between horns and the C-shape of the lateral ventricle. Since the three skeletons encounter each other at the atrium center, the *atrium width* is estimated by averaging the width at the atrium center for the three horn skeletons. We define a pair of angle measures – *starting and ending angles* – with respect to two neighboring horns. The starting and ending angles are the angles between the starting and ending vectors of two horns, respectively. The starting angle estimates the degree of the relative orientations of each horn nearby the atrium, and the ending angle quantifies the angle between the tip of the horns with respect to the atrium center.

4   Morphology Analysis Using the Feature Descriptors

In order to assess the descriptive ability of the feature descriptors, we performed two analyses on a dataset of a study of aging: (1) a feature-based description of the shape variations of the lateral ventricle within a population, and (2) a ventricular enlargement analysis using the feature descriptors in relation to general brain atrophy. Without loss of generality, we present the experiment results of the left lateral ventricle only. The morphology of the right lateral ventricle can be quantified in the same way.

The dataset includes T1-weighted MR images obtained at a GE Signa HDxt 1.5T scanner, (General Electric, USA) from 33 participants (10 women and 23 men, age $= 72.7 \pm 0.7$ years) randomly selected from The Lothian Birth Cohort 1936 study [6]. The scanning protocol is described in detail elsewhere [21]. The dataset also contains the brain tissue volume (BTV) and intracranial volume, and the binary masks of the lateral ventricles, following the segmentation procedure described in [3]. The results were visually assessed by a trained image analyst and manually corrected. From these binary ventricular masks we obtained the surface models as described previously. We validated the accuracy of this process using the Dice coefficient and the symmetric mean distance between the models and the target volumes. The reconstructed models showed high accuracy with the segmentations: Dice coefficient was 0.954 ± 0.013, and the mean distance was 0.642 ± 0.763 mm for all subjects.

4.1   Feature-Based Description of Shape Variations across all Subjects

In order to investigate the shape variations of the lateral ventricles across the sample, we first computed the mean surface shape and its deviations using principal component analysis (PCA) and transformed them into the feature space. For the PCA, we normalized the surface models via isotropic rescaling using the intracranial volume, and aligned them optimally by matching the atrium center and rotating each of them to minimize the between-surface distance via a Procrustes analysis [9]. The mean surface model was computed simply by averaging the corresponding points on the surface models: $\bar{x} = \sum_{i=1}^{n} x_i$, where x_i is a $3 \times k$ vector describing the surface model with k points and n is the number of subjects. The covariance matrix is given by: $D = \frac{1}{n-1} \sum_{i=1}^{n} (x_i - \bar{x}) \cdot (x_i - \bar{x})^T$. The eigen-decomposition on D delivers the $\min(n-1, 3k)$ principal modes of the variation [11]. A mode with a high variance (i.e. large eigenvalue of D) represents a larger part of the shape variation of the lateral ventricle across subjects. The percentage of the shape variability of each mode is determined by the ratio between the corresponding eigenvalue and the sum of the eigenvalues [16]. We generated the surface models showing the shape variations of each mode between ± 3 standard deviations (SD) and extracted the feature descriptors from them.

Figure 2 shows the surface meshes of the largest and smallest size with the average surface mesh and the surface models representing the shape variation of the first and second modes between ± 3 SD with their skeletons. The percentage of the shape variability of the first mode was 50.75 % in this population, and the second mode was 13.62 %. The shape variability of each mode can be explained more intuitively using the feature descriptors. For example, the visual observation of the morphological variation along the 'frontal horn' of ± 3 SD of the first mode in Figure 2 can be described based on the measured values of feature descriptors (Table 1) as follows: the average width, starting width, and the geodesic length of the frontal horn decreased, while the bending increased from the -3 to the +3 SD. Moreover, the morphological relationship between two

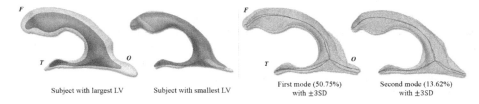

| Subject with largest LV | Subject with smallest LV | First mode (50.75%) with ±3SD | Second mode (13.62%) with ±3SD |

Fig. 2. (Left) ventricular surface meshes (white) of the largest and smallest size with the average surface mesh (red). (Right) Surface meshes with their skeletons representing the shape variation of the first and second modes between ± 3 standard deviations (SD). The surface, skeleton and bars on blue color correspond to + 3 SD, and those on green color to − 3 SD. F: frontal horn, T: temporal horn and O: occipital horn.

horns is expressed by the inter-horn feature descriptors. For instance, the bending of the temporal and occipital horns highly increased towards the -3 SD. On the other hand, the start and ending angles between these two horns increased on the +3 SD surface model, which reflected the two horns were widely opened than -3 SD. Therefore, analyzing both sets of feature descriptors together we can interpret that the temporal and occipital horn are bending to each other when ventricular shapes differ from the mean going from +3 to -3 SD of the first mode. Observe on Figure 2 that the other values of the feature descriptors were also reflecting consistently the shape differences between the surface models of ± 3 SD of the first mode.

Table 1. Measure values of the feature descriptors on the surface meshes of ± 3 SD of the first mode

| | Individual-horn Feature Descriptors | | | | | | | | | | | |
| | Frontal Horn | | | | Temporal Horn | | | | Occipital Horn | | | |
	MW	SW	GL	Bend	MW	SW	GL	Bend	MW	SW	GL	Bend
+3SD	4.71	5.16	90.43	0.20	1.50	4.60	51.89	0.13	2.55	5.18	23.01	0.05
Avg	4.67	4.28	89.24	0.21	1.77	4.18	50.67	0.11	2.23	4.74	22.42	0.03
-3SD	4.65	4.29	88.56	0.22	2.00	4.47	49.29	0.10	2.39	4.00	22.86	0.02

| | Inter-horn Feature Descriptors | | | | | |
	AW	SA (FT)	SA (TO)	SA (OF)	EA (FT)	EA (TO)	EA (OF)
+3SD	6.11	108.19	102.10	149.22	33.05	133.80	200.13
Avg	5.92	102.78	112.43	144.65	32.88	137.83	198.67
-3SD	5.59	112.61	120.92	126.13	33.08	143.81	162.52

MW.: mean width (mm), SW.: starting width (mm), GL.: geodesic length (mm), Bend.: bending, AW.: atrium width (mm), SA.: starting angle (°), EA.: ending angle (°), FT.:frontal-temporal, TO.: temporal-occipital, and OF.: occipital-frontal

±3SD: surface mesh corresponding to ±3 SD of the first mode, and Avg.: average surface mesh

4.2   A Morphological Feature Analysis with a General Brain Atrophy Measure

It is known that cerebral atrophy is a common accompaniment of aging and manifests as decreased total brain volume and increased ventricular volumes [3]. In order to evaluate the descriptive ability of the feature descriptors in relation to brain atrophy, we examined the associations between each feature descriptor and brain tissue volume (BTV). For this we performed a robust multilinear regression using the function "robustfit" from MATLAB Statistical Toolbox. The regression model includes the feature descriptors as the dependent variable and BTV as the independent variable, with gender and age as covariates.

The regression analysis showed that the average width of each horn was increased as the BTV decreased (β = -8.197 ~ -10.470, P < 0.05). The width of the atrium also increased as the BTV decreased (β = -20.652, P < 0.05). Near the atrium of the lateral ventricle, the width of the frontal and occipital horns were not significantly associated with brain atrophy (P > 0.05), but it was for the temporal horn (β = -17.791, P < 0.05). The geodesic length and bending of each horn were not significantly related to brain atrophy and the angular measurements between horns with respect to the atrium center were not associated to the brain atrophy either. These results indicate that the association between ventricular enlargement and general brain atrophy is manifested as the significant increase of the width along the ventricular central axes (i.e. skeleton) despite of the complex patterns of shape variations observed in this population sample through PCA.

Table 2. Association between a brain atrophy measure (brain tissue volume) and the feature descriptors of the lateral ventricle using a robust multilinear regression (beta, P).

Linear Regression on Individual Horn Structural Feature Descriptors				
	Mean Width	Starting Width	Geodesic Length	Bending
Frontal Skeleton	**-10.470, 0.011**	-12.552, 0.100	28.109, 0.328	0.283, 0.068
Temporal Skeleton	**-8.197, 0.002**	**-17.791, 0.003**	-4.239, 0.849	-0.043, 0.820
Occipital Skeleton	**-8.763, 0.039**	-9.820, 0.104	-60.131, 0.370	0.165, 0.132

Linear Regression on Inter-horn Structural Feature Descriptors	
Atrium Width	**-20.652, 0.035**
	Angle between Starting Points with Atrium Center	Angle between Ending Points with Atrium Center
Frontal-Temporal	18.224, 0.749	-21.239, 0.159
Temporal-Occipital	-11.750, 0.826	23.766, 0.501
Occipital-Frontal	-22.718, 0.603	-18.809, 0.871

Regression Model: Structural Feature (Enlargement) = β_1* Brain Tissue Volume/Intracranial Volume + β_2 * Gender + β_3 * age
beta, P : Unstandardized coefficients and P-values of the regression models
Significant values (P < 0.05) are highlighted as boldface

5 Conclusion

In this paper, we introduce a set of descriptors, that characterize the lateral ventricle's morphology, with an explicit measurement basis that includes the definitions of the ventricular skeleton and atrium center on the shape model. We demonstrate through statistical analyses on 33 randomly selected datasets from a study of aging how the feature descriptors can express the changes of the lateral ventricle with respect to the general brain atrophy. We also present the descriptive ability of the feature descriptors using PCA-based shape models representing the shape variations across subjects. The structural feature descriptors can provide a precise description of the ventricular shapes based on anatomical knowledge for the morphology analysis of the lateral ventricle. We anticipate that these shape feature descriptors of the lateral ventricle would be useful in characterizing its deformation more systematically. Further work is needed to investigate the relationships between each feature descriptor and the role they play in the volumetric changes observed on the lateral ventricles. In addition, the application to larger samples is also needed to generalize the sensitivity of the feature descriptors and to validate the consistency of the anatomical landmarks (e.g. tips of the ventricular horns) across subjects.

Acknowledgments. This work was funded by the National Research Foundation of Korea (Grant no. 2012K2A1A2033133/no.2011-0009761), the Row Fogo Charitable Trust, the Scottish Imaging Network A Platform for Scientific Excellence and Age UK for the LBC1936 Study. We also thank the Lothian Birth Cohort 1936 Study Collaborative Group at The University of Edinburgh led by Profs. Ian J. Deary and Joanna M. Wardlaw, who provided the data used in this manuscript.

References

1. Antiga, L., Ene-Iordache, B., Remuzzi, A.: Computational geometry for patient-specific reconstruction and meshing of blood vessels from MR and CT angiography. IEEE Trans. Med. Imaging 22(5), 674–684 (2003)
2. Apostolova, L.G., Green, A.E., Babakchanian, S., Hwang, K.S., Chou, Y.Y., Toga, A.W., Thompson, P.M.: Hippocampal atrophy and ventricular enlargement in normal aging, mild cognitive impairment (MCI), and Alzheimer Disease. Alz. Dis. Assoc. Dis. 26(1), 17–27 (2012)
3. Aribisala, B.S., Valdés Hernández, M.C., Royle, N.A., Morris, Z., Muñoz Maniega, S., Bastin, M.E., Deary, I.J., Wardlaw, J.M.: Brain atrophy associations with white matter lesions in the ageing brain: the Lothian Birth Cohort 1936. Eur. Radiol. 23(4), 1084–1092 (2013)
4. Bourne, S.K., Conrad, A., Konrad, P.E., Neimat, J.S., Davis, T.L.: Ventricular width and complicated recovery following deep brain stimulation surgery. Stereotact Funct. Neurosurg. 90(3), 167–172 (2012)
5. Buckner, R.L.: Memory and executive function in aging and ad: multiple factors that cause decline and reserve factors that compensate. Neuron 44(1), 195–208 (2004)

6. Deary, I.J., Gow, A.J., Taylor, M.D., Corley, J., Brett, C., Wilson, V., Campbell, H., Whalley, L.J., Visscher, P.M., Porteous, D.J., Starr, J.M.: The Lothian Birth Cohort 1936: a study to examine influences on cognitive ageing from age 11 to age 70 and beyond. BMC Geriatr. 7, 28 (2007)
7. Ferrarini, L., Palm, W.M., Olofsen, H., van Buchem, M.A., Reiber, J.H.C., Admiraal-Behloul, F.: Shape differences of the brain ventricles in Alzheimer's disease. NeuroImage 32(3), 1060–1069 (2006)
8. Frisoni, G.B., Geroldi, C., Beltramello, A., Bianchetti, A., Binetti, G., Bordiga, G., DeCarli, C., Laakso, M.P., Soininen, H., Testa, C., et al.: Radial width of the temporal horn: a sensitive measure in alzheimer disease. American Journal of Neuroradiology 23(1), 35–47 (2002)
9. Gower, J.C.: Generalized procrustes analysis. Psychometrika 40(1), 33–51 (1975)
10. Gutman, B.A., Wang, Y., Rajagopalan, P., Toga, A.W., Thompson, P.M.: Shape matching with medial curves and 1-D group-wise registration. In: 2012 9th IEEE International Symposium on Biomedical Imaging, pp. 716–719 (2012)
11. Heimann, T., Meinzer, H.P.: Statistical shape models for 3D medical image segmentation: A review. Med. Image Anal. 13(4), 543–563 (2009)
12. Heitz, G., Rohlfing, T., Maurer, J.C.R.: Statistical shape model generation using nonrigid deformation of a template mesh. In: SPIE on Medical Imaging, pp. 1411–1421 (2005)
13. Jeong, H.S., Lee, S., Yoon, S., Jung, J.J., Cho, H.B., Kim, B.N., Ma, J., Ko, E., Im, J.J., Ban, S., Renshaw, P.F., Lyoo, I.K.: Morphometric abnormalities of the lateral ventricles in methamphetamine-dependent subjects. Drug Alcohol Depend 131(3), 222–229 (2013)
14. Kim, J., Park, J.: Organ Shape Modeling Based on the Laplacian Deformation Framework for Surface-Based Morphometry Studies. J. Comp. Sci. Eng. 6, 219–226 (2012)
15. Lee, J.H., Yoon, S., Renshaw, P.F., Kim, T.S., Jung, J.J., Choi, Y., Kim, B.N., Jacobson, A.M., Lyoo, I.K.: Morphometric changes in lateral ventricles of patients with recent-onset type 2 diabetes mellitus. PloS One 8(4), e60515 (2013)
16. Lu, Y.C., Untaroiu, C.D.: Statistical shape analysis of clavicular cortical bone with applications to the development of mean and boundary shape models. Comput. Meth. Prog. Bio. 111(3), 613–628 (2013)
17. Ng, H.-F., Chuang, C.-H., Hsu, C.-H.: Extraction and Analysis of Structural Features of Lateral Ventricle in Brain Medical Images. In: 2012 Sixth International Conference on Genetic and Evolutionary Computing (ICGEC), pp. 35–38 (2012)
18. Nolte, J.: Essentials of the Human Brain. Elsevier Health Sciences (2009)
19. Piccinelli, M., Veneziani, A., Steinman, D.A., Remuzzi, A., Antiga, L.: A framework for geometric analysis of vascular structures: application to cerebral aneurysms. IEEE Trans. Med. Imaging 28(8), 1141–1155 (2009)
20. Wan, K.R., Toy, J.A., Wolfe, R., Danks, A.: Factors affecting the accuracy of ventricular catheter placement. J. Clin. Neurosci. 18(4), 485–488 (2011)
21. Wardlaw, J.M., Bastin, M.E., Valdés Hernández, M.C., Maniega, S.M., Royle, N.A., Morris, Z., Clayden, J.D., Sandeman, E.M., Eadie, E., Murray, C., Starr, J.M., Deary, I.J.: Brain aging, cognition in youth and old age and vascular disease in the Lothian Birth Cohort 1936: rationale, design and methodology of the imaging protocol. Int. J. Stroke 6(6), 547–559 (2011)

Extension of an MRI Simulator Software for Phase Contrast Angiography Experiments[*]

Alexandre Fortin[1,**], Emmanuel Durand[2], and Stéphanie Salmon[1,**]

[1] Université de Reims Champagne-Ardenne, LMR, France
alexandre.fortin@etudiant.univ-reims.fr,
stephanie.salmon@univ-reims.fr
[2] Université de Strasbourg, CNRS, ICube & HUS, France
http://dl.free.fr/vsXyJtBaM

Abstract. The purpose of our work is to develop tools for simulation of angiographic MRI images of cerebral vasculature. We present here our extension of the open-source software Jemris dedicated to this goal. Jemris is an advanced simulation software including most of MRI physical phenomena (concomitant gradient fields, molecular diffusion, patient move, etc). Our work now provides the additional ability to simulate fluids in motion, in addition to static tissues. These changes allow us to obtain velocimetric phase contrast images for simple Poiseuille flow, with multi-directional speed encoding. Comparison between our simulations and real MRI data gives promising first results.

1 Introduction

This work is part of a global project whose aim is to get a collection of softwares dedicated to the simulation of virtual angiography using 3D+t vascular models. In other words, first we reconstruct a given patient's vasculature from MRI images of his brain and compute the blood flow in this realistic geometry. Then, we generate virtual angiographic MRI images from these calculated flow data, in order to validate the models used, by comparison with real patient MRI images. It uses techniques of image segmentation, 3D vascular geometry models, computational fluid dynamics and MRI simulation. We focus here on MRI angiographic simulation, carried out by extending Jemris software.

2 Extension of an MRI Simulator

Jemris is an advanced MRI simulator software written in C++, open-source, extensible, and freely modifiable [1]. Most physical MRI phenomena are considered, such as non-uniform gradients, concomitant fields [2], noise, patient movements, molecular diffusion, etc. As most of similar softwares [3] [4] [5], Jemris is based on isochromats summation. It involves splitting the object into elementary volumes (called "spins" for convenience) and summing magnetization signal generated by each one.

---

[*] This research was funded by *Agence Nationale de la Recherche* (Grant Agreement ANR-12-MONU-0010).
[**] Corresponding author.

F. Bello and S. Cotin (Eds.): ISBMS 2014, LNCS 8789, pp. 150–154, 2014.
© Springer International Publishing Switzerland 2014

Fig. 1. Our modifications allow Jemris to simulate individual trajectories for each spin. Here an example of simple gradient echo MRI sequence with two static spins at the top and two moving spins at the bottom.

Our work consists of providing an extension to the C++ code, in order to simulate fluid traveling. Movements in Jemris are based on a Lagrangian approach, which means to know individual spins trajectories. Magnetization signal is determined by solving numerically Bloch equations for each spin, depending on the external magnetic field value. Varying spin position over time changes the field value seen by the particle:

$$\mathbf{r} = \mathbf{r}(t) \quad \Rightarrow \quad \mathbf{B}(\mathbf{r}, t) = \mathbf{B_0} + \mathbf{G}(t).\mathbf{r}(t)$$

$\mathbf{B}(\mathbf{r}, t)$ is the total external magnetic field, with $\mathbf{B_0}$ the main field and $\mathbf{G}(t)$ the field gradients used in MRI for spatial localization.

However, by default, Jemris only authorizes to specify one trajectory for all spins, in order to simulate movement of a rigid sample. So we added a specific class to the C++ code to allow users to specify a different trajectory for each spin. A simple example of MRI simulation performed with our modified version of Jemris is shown on Fig. 1. As it appears on that image, we used four spins with specific individual behaviors: two static spins (at the top of the image), and two moving spins with diagonal trajectories (at the bottom). With this new version, it thus becomes possible to describe flow phenomena, as we expect.

3   Simulation of Phase Contrast Velocimetry

3.1   Discretized Poiseuille Flow

Our first simulations are performed on synthetic input data of discretized Poiseuille flow. We determine each individual spin trajectory based on the general Poiseuille law. It means, for the ith spin, with horizontal move along z axis:

$$z_i(t) = z_i(0) + V_{max}.\left(1 - \frac{(r_i - r0)^2}{R^2}\right).t$$

r_i, z_i are the cylindrical coordinates of the ith spin, r0 the coordinate of the center of the tube, R its radius and V_{max} the maximum velocity.

3.2   Phase Contrast Simulation and Data Treatment

Phase contrast is a well-known non-invasive angiographic MRI technique. The magnetization phase shift of the flowing spins becomes proportional to their displacement. It allows to carry out some velocimetric measurements on each pixel of the image [6].

We perform here 1D-velocity encoding by orienting the flowing spins along frequency encoding direction (in MRI, spatial localization of the tissues is made possible by gradients field, with a first encoding direction linked to spins frequency and a second to their phase). In order to eliminate static structures, we carry out two acquisitions with opposite velocity encoding. After Fourier transform of raw data, we obtain velocity maps by complex division of both acquisitions [7].

4   Comparison with Experimental Data

We perform our experiences with a 3 Tesla MRI machine. Our physical phantom is a flexible tube with flowing water, itself immersed in static water. A system of pump and overflowed vases ensures a constant flow rate. A pulsatile mode is also intended to reproduce heartbeat, and the compliance of the tube is supposed to imitate that of arteries [8], in expectation of future comparisons with cerebral vasculature.

5   Results

The Poiseuille simulation presented here, with about 20 000 spins, including flowing and static spins, took about one hour per acquisition on one CPU, for short MRI phase contrast sequences of 32s. Comparison with experimental data and theoretical Poiseuille map are shown on Fig. 2 and Fig. 3.

6   Discussion and Prospect

The validation of simulated data by experimental ones is, at the moment, limited because of differences in physical input conditions. We should obtain some results with parameters closer to experimental data. It will imply to increase spatial resolution, sample dimensions and fluid velocity, which will also increase the number of spins. This will maybe lead us to parallelize our simulations on multiples CPU to reduce time computation. Some coupling methods of Computational Fluid Dynamics and angiographic simulation already exist [9] [10] [11] [12] [13], but performing our flow simulations with an advanced and independent MRI software allows us to obtain a good quality of images at lower cost, with possible extension to arbitrary complex flow geometries.

Our next step will be to simulate phase images with flow data from simulated vasculature, *i.e.* from spins trajectories based on computational fluid dynamics. Simulation of a full brain, for comparison with real patients angiographic MRI data would be ideal. However, it seems to be difficult in regards of time calculation and input data amount (400 To estimated). Subsequent studies will discuss what proportion of vasculature it is reasonable to simulate.

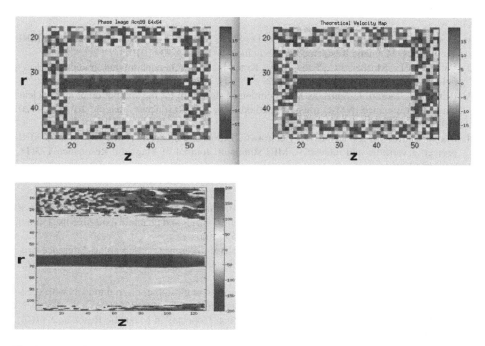

Fig. 2. Top: Left: Simulated phase image of Poiseuille flow with 23 757 spins, maximum velocity of 15 mm/s, spatial resolution of 25/100 mm^{-1} and 1% of random noise. Static spins are present on both sides of the flow. Right: "Perfect" theoretical velocity map of Poiseuille flow. Bottom: Experimental phase image of our physical phantom with maximum velocity of 167 mm/s and spatial resolution of 64/100 mm^{-1}. On each image, the blue rectangle corresponds to Poiseuille flow and the green ones to static structures. The rest is noise due to air.

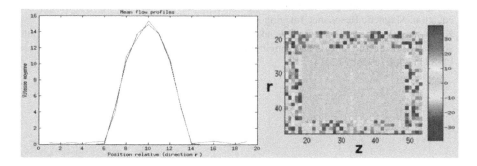

Fig. 3. Left: Superposition of Poiseuille theoretical (blue) and simulated (green) mean velocity profiles. Right: Difference between theoretical velocity map and simulated one. Differences are close to zero in the central rectangle (we observe a standard deviation of 1.2 mm/s). Future comparison with experimental velocity data will be interesting but is, at the moment, unsignificant because of the differences of size, spatial resolution and flow rate.

References

1. Stocker, T., Vahedipour, K., Pflugfelder, D., Shah, N.J.: High-performance computing MRI simulations. Magnetic Resonance in Medicine 64(1), 186–193 (2010)
2. Volegov, P.L., Mosher, J.C., Espy, M.A., Kraus Jr., R.H.: On concomitant gradients in low-field MRI. Journal of Magnetic Resonance 175(1), 103–113 (2005)
3. Bittoun, J., Taquin, J., Sauzade, M.: A computer algorithm for the simulation of any nuclear magnetic resonance (NMR) imaging method. Magnetic Resonance Imaging 2(2), 113–120 (1984)
4. Benoit-Cattin, H., Collewet, G., Belaroussi, B., Saint-Jalmes, H., Odet, C.: The SIMRI project: a versatile and interactive MRI simulator. Journal of Magnetic Resonance 173(1), 97–115 (2005)
5. Jochimsen, T.H., von Mengershausen, M.: ODIN: Object-oriented development interface for NMR. Journal of Magnetic Resonance 170(1), 67–78 (2004)
6. Durand, E.P., Jolivet, O., Itti, E., Tasu, J.P., Bittoun, J.: Precision of magnetic resonance velocity and acceleration measurements: theoretical issues and phantom experiments. Journal of Magnetic Resonance Imaging 13(3), 445–451 (2001)
7. Bernstein, M.A., Ikezaki, Y.: Comparison of phase-difference and complex-difference processing in phase-contrast MR angiography. Journal of Magnetic Resonance Imaging 1(6), 725–729 (1991)
8. Stevanov, M., Baruthio, J., Eclancher, B.: Fabrication of elastomer arterial models with specified compliance. Journal of Applied Physiology 88(4), 1291–1294 (2000)
9. Marshall, I.: Simulation of in-plane flow imaging. Concepts in Magnetic Resonance 11(6), 379–392 (1999)
10. Watanabe, M., Kikinis, R., Westin, C.F.: Level set-based integration of segmentation and computational fluid dynamics for flow correction in phase contrast angiography. Academic Radiology 10(12), 1416–1423 (2003)
11. Ford, M., Stuhne, G., Nikolov, H., Habets, D., Lownie, S., Holdsworth, D., Steinman, D.: Virtual angiography for visualization and validation of computational models of aneurysm hemodynamics. IEEE Transactions on Medical Imaging 24(12), 1586–1592 (2005)
12. Jurczuk, K., Kretowski, M., Bellanger, J.J., Eliat, P.A., Saint-Jalmes, H., Bezy-Wendling, J.: Computational modeling of MR flow imaging by the lattice Boltzmann method and Bloch equation. Magnetic Resonance Imaging 31(7), 1163–1173 (2013)
13. Klepaczko, A., Szczypinski, P., Dwojakowski, G., Strzelecki, M., Materka, A.: Computer simulation of magnetic resonance angiography imaging: model description and validation. PloS One 9(4), e93689 (2014)

Preliminary Study on Finite Element Simulation for Optimizing Acetabulum Reorientation after Periacetabular Osteotomy

Li Liu[1], Timo Michael Ecker[2], Steffen Schumann[1], Klaus Siebenrock[2], and Guoyan Zheng[1]

[1] Institute for Surgical Technology and Biomechanics,
University of Bern, Switzerland
{li.liu,guoyan.zheng}@istb.unibe.ch
[2] Orthopaedic Department, University of Bern, Switzerland

Abstract. Periacetabular osteotomy (PAO) is an effective approach for surgical treatment of hip dysplasia. The aim of PAO is to increase acetabular coverage of the femoral head and to reduce contact pressures by reorienting the acetabulum fragment after PAO. The success of PAO significantly depends on the surgeon's experience. Previously, we have developed a computer-assisted planning and navigation system for PAO, which allows for not only quantifying the 3D hip morphology for a computer-assisted diagnosis of hip dysplasia but also a virtual PAO surgical planning and simulation. In this paper, based on this previously developed PAO planning and navigation system, we developed a 3D finite element (FE) model to investigate the optimal acetabulum reorientation after PAO. Our experimental results showed that an optimal position of the acetabulum can be achieved that maximizes contact area and at the same time minimizes peak contact pressure in pelvic and femoral cartilages. In conclusion, our computer-assisted planning and navigation system with FE modeling can be a promising tool to determine the optimal PAO planning strategy.

Keywords: Periacetabular osteotomy, computer-assisted diagnosis and planning, finite element simulation.

1 Introduction

Periacetabular osteotomy (PAO) is an effective approach for surgical treatment of hip dysplasia in young patients [1]. The aim of PAO is to increase acetabular coverage of the femoral head and to reduce contact pressures by realigning the hip joint [2]. It was reported that PAO planning approach is mainly based on two types of optimization strategies which are morphology-based and biomechanics-based optimization, respectively. In clinical routine, diagnosis and pre-operative planning of hip dysplasia is based on hip joint morphological parameters measured from a plain radiograph. However, these radiographic parameters for diagnosing hip dysplasia are unreliable. For instance Clohisy et al. [3] evaluated the reliability of six hip specialists identifying important radiographic features

F. Bello and S. Cotin (Eds.): ISBMS 2014, LNCS 8789, pp. 155–162, 2014.

of the hip on plain radiographs. They concluded that the standard radiographic parameters used to diagnose dysplasia are not reproducible [3]. Additionally, the same group, Carlisle et al. [4] further investigated the reliability of radiographic measurements of the hip by various musculoskeletal physicians. They found that while the measurements were reliable for a given observer, the measurements were less reliable across observers and were limited in determining a consistent radiographic diagnosis.

The other type of planning strategy is based on biomechanics optimization. Zhao et al. [5] conducted a 3D finite element (FE) analysis of acetabular dysplasia. The effects of dysplasia and PAO were both investigated by analyzing the change in Von Mises stress in the cortical bone before and after surgery. They showed that the PAO may be beneficial. One limitation of this method lies in that the acetabular dysplastic model representing different levels of severity of dysplasia were generated by deforming the acetabular rim of a normal hip. Thus, it ignores the influence of the abnormal shapes of the femoral head and the acetabulum of the real dysplastic hip. In contrast, the computer-assisted Biomechanical Guidance System (BGS) introduced by Armand et al. [6] combines geometric and biomechanical feedback with intra-operative tracking to guide the surgeon through the PAO procedure. During the planning stage, the PAO planning computes contact pressures via Discrete Element Analysis (DEA) in order to suggest a reorientation of the acetabulum that minimizes simultaneous peak contact pressure in sitting, standing, and walking positions [7]. Recently, Zou et al. [8] developed a 3D FE simulation of PAO in 5 models generated from CT scans of dysplastic hips. The acetabulum of each model was rotated in 5° increments in the coronal plane from original Lateral Center Edge (LCE) angle, and the relationship between contact area and pressure and Von Mises stress in the femoral and pelvic cartilage were investigated until the optimal position for the acetabulum following PAO was found.

Previously, we have developed a computer-assisted planning and navigation system for PAO [9]. This comprehensive system allows for not only quantifying the 3D hip morphology with geometric parameters such as acetabular orientation (expressed as inclination and anteversion angles with respect to the so-called Anterior Pelvic Plane (APP) [10]), LCE angle and acetabular coverage ratio (AC) for a computer-assisted diagnosis of hip dysplasia but also virtual PAO surgical planning and simulation (Fig. 1). In this paper, based on this previously developed PAO planning and navigation system, we developed a 3D FE model based on patient-specific geometry to investigate the optimal acetabulum reorientation after PAO. Our aim is to develop a biomechanical tool to determine the optimal PAO planning strategy.

2   Materials and Methods

2.1   Biomechanical Model Based on Patient-Specific Geometry

Mesh Generation. Bone surface models of the reoriented hip joints are imported into ScanIP software (Simpleware Ltd, Exeter, UK) as shown in Fig.

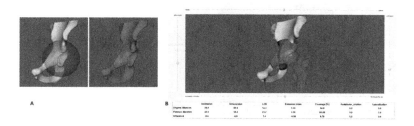

Fig. 1. Virtual PAO surgery in our computer-assisted PAO planning system. (A) Virtual cutting operation is done with a sphere, whose radius and position can be interactively adjusted; (B) Virtual reorientation operation is done by interactively reorienting the acetabular fragment around the hip center. During the reorientation, real-time computation of the inclination and anteversion angles [10] are provided as a feedback to the surgeon.

2(A) and (C). While patient-specific cartilage models are essential in biomechanical simulation, it has been previously reported in [11] that the predicted optimal alignment of the acetabulum was not significantly sensitive to the choice of cartilage thickness distribution over the acetabulum. Therefore a 3D dilation operation was performed on femoral head and actabular surfaces to create femoral and pelvic cartilage layers with a constant thickness as has been done in [12]. Surfaces were discretized using tetrahedral elements (Fig. 2(B) and (D)). Since the primary concern was focused on the joint contact, a finer mesh was employed for the cartilage than for the bone. Refined tetrahedra meshes (218316 elements) were constructed for the cartilage layers using ScanFE module (Simpleware Ltd, Exeter, UK). Cortical bone surfaces were discretized using coarse tetrahedra elements (307419 elements). Trabecular bone was not included in the models, for it has little effect on predictions of contact stress as reported in [12].

Material Property. Pelvic and femoral cartilages were modeled as homogeneous, isotropic, and linearly elastic material with Young's Modulus E = 15 MPa and Poissons ratio = 0.45. Cortical bone of pelvis and femur were modeled as homogeneous, isotropic material with elastic modulus E = 17 GPa and Poissons ratio = 0.3 as has been suggested in [8].

Boundary Conditions and Loading. Tied and sliding contact constraints were used in Abaqus/CAE 6.10 (Dassault Systmes Simulia Corp, Providence, RI, USA) to define the cartilage-to-bone and cartilage-to-cartilage interfaces, respectively. It has been reported in [13] that the friction coefficient between articulating cartilage surfaces is very low, on the order of 0.01-0.02 in the presence of synovial fluid. Therefore, it is reasonable to neglect frictional shear stresses between contacting articular surfaces. The loading and boundary conditions used in this paper resemble those used by Phillips et al. [15] (Fig. 2 (E)). The top surface of pelvis and pubic areas were fixed, and the distal end of the femur was constrained to prevent displacement in the body x and y directions while being free in vertical z direction (Fig. 2 (E)). The center of femoral head derived from

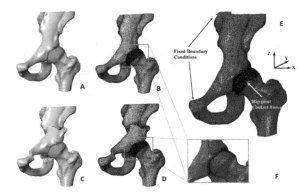

Fig. 2. Biomechanical simulation of contact pressure on acetabular cartilage. (A) Surface models; (B) Volume meshes; (C) Surface models after acetabulum fragment reorientation; (D) Volume meshes for the reoriented situation; (E) Boundary conditions and loading for biomechanical simulation; (F) Coarse meshes for pelvic and femoral models, and refined meshes for pelvic and femoral cartilages.

a least-square sphere fitting was selected to be the reference node. The nodes of femoral head surface are constrained by the reference node via kinematic coupling. The fixed boundary condition model was then subjected to a loading condition as published in [14], representing a single leg stance situation with the resultant hip joint contact force acting at the reference node. The loading condition consists of static and dynamic loading scenarios. In static loading scenario, constant components of contact force were applied to hip joint. Following the loading specification in [15], the components of joint contact force along 3 axes are given as 195N, 92N, and 1490N, respectively, by assuming a constant body weight of 650N for all subjects to remove any scaling effect of body weight on the absolute value of the contact pressure. The resultant force is applied based on anatomical coordinate system described in Bergmann et al [14], whose local coordinate is defined with the x axis running between the centers of the femoral heads (positive running from the left femoral head to the right femoral head), the y axis pointing directly anteriorly, and the z axis pointing directly superiorly. However, in dynamic loading scenario, a time-dependent force function with duration of 11 seconds was applied to the center of hip joint. This dynamic time-dependent force function was derived from the in-vivo measurement data published by Bergmann et al. [14]. The mesh models of original dysplastic hip and a series of planned situations were imported into Abaqus simulation environment for biomechanical simulation (see Fig. 2 for details).

2.2   Experiment Design and Results

A preliminary study was conducted on CT scans of two patients with hip dysplasia in order to verify the efficacy of the developed FE model. Each patient

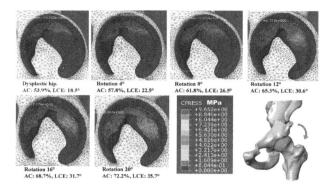

Fig. 3. Contact pressure distribution on acetabular cartilage for the 1st dysplastic hip in static one-leg stance situation. The acetabulum fragment is virtually rotated in $4°$ increment in the APP from the original acetabulum inclination angle.

scan was performed in the supine orientation and covered the entire joint region (the acetabulum and proximal femur) in axial slices. The predicted peak contact pressure and total contact area are directly extracted from the output of Abaqus. We then compared quantitatively the peak contact pressure and contact area on acetabulum cartilage in different acetabulum position and investigated optimal planning strategy in static one-leg stance loading scenario.

Fig. 3 shows how contact pressure distribution of the pelvic cartilage changed for the 1st dysplastic hip when AC and 3D LCE angle were increased. The contact area originally focused on the superior region and almost no contact area was in the anterior and posterior regions. When the AC and 3D LCE angle were increased, the contact area shifted from lateral region towards the medial region. Fig. 4 (A) and (C) present peak contact pressures at different AC and 3D LCE angles. Contact areas are shown in Fig. 4 (B) and (D) as well. The peak contact pressures and contact areas are available for both pelvic and femoral cartilages, but in our study only contact pressure pattern of the pelvic cartilage was investigated. An optimal acetabulum fragment reposition with minimum peak contact pressure and maximum contact area was achieved for both dysplastic hips. More importantly, for each hip, both the minimal peak contact pressure and the maximum contact area were achieved at the same acetabulum fragment reposition. A large rotation of acetabulum does not guarantee low peak contact pressure and large contact area. For two dysplastic hips, we found that the peak contact pressures on the optimal planned situations are 6.3MPa and 3.8MPa, respectively, while their contact areas are quite different (833.9mm^2 and 1496.7mm^2, respectively). The reason is simply that these two patients have different acetabulum. For example, the diameters of acetabulum rims from these two patients are 43.6mm and 52.6mm, respectively.

When the optimized position of an acetabulum was found, we then compared the peak contact pressure and the contact area between the original dysplastic hip and the optimally reoriented hip in dynamic one-leg stance loading condition

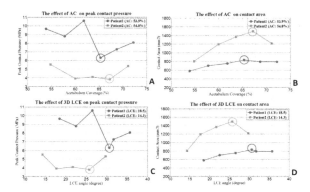

Fig. 4. (A) Effect of AC on hip joint peak contact pressure. Circled points represent the lowest pressures for each subject; (B) Effect of AC on hip joint contact area. Circled points indicate the largest contact areas for each subject; (C) Effect of 3D LCE on hip joint peak contact pressure. Circled points represent the lowest pressures for each subject; (D) Effect of 3D LCE on hip joint contact area. Circled points indicate the largest contact areas for each subject.

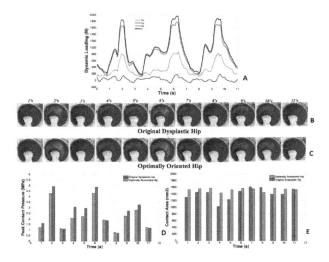

Fig. 5. Contact pressure distribution on acetabular cartilages for dynamic one-leg stance loading scenario. (A) Time-dependent dynamic loading input; (B) Contact pressure distribution of the original dysplastic hip over the 11 seconds simulation; (C) Contact pressure distribution of the optimally reoriented hip over the same 11 seconds simulation; (D) Comparison of peak contact pressures between the original dysplastic hip and the optimally reoriented hip; (E) Comparison of contact areas between the original dysplastic hip and the optimally reoriented hip.

where a time-dependent hip joint force function was applied [14]. The simulation process lasted 11 seconds in total, and the time-dependent force curve was shown in Fig. 5(A). We compared the difference of corresponding peak contact pressure and contact area at each second (see Fig. 5(B) and (C)) between the original dysplastic hip and the optimally reoriented hip. Our simulation results show that when the results of the optimally reoriented hip were compared with those of the original dysplastic hip, the peak pressures were reduced while the contact areas were increased (see Fig .5(D) and (E)).

3   Discussions and Conclusion

In this paper, we proposed and developed a 3D FE Model based on patient-specific geometry for optimization of the acetabulum reorientation after PAO. A preliminary study conducted on two dysplastic hips were used to verify the efficacy of the developed FE model. Both static and dynamic loading conditions were investigated. Although CT scan was performed in the supine position and the loading condition of our biomechanical simulation is based on one-leg stance situation [14], previous work [11] has shown that there is no significant difference between the contact pressure in the one-leg stance reference frame and those in the supine reference frame. In addition, as pointed out by Armiger et al. [7], it is not an infrequent clinical practice to use models derived from the supine frame to do biomechanical simulation of the standing frame. Therefore we believe our model makes good use of valuable, available data from the original Bergmann's work [14]. Compared to the results reported by Zou et al. [8], who also developed a 3D FE simulation of PAO in order to find optimal reorientation position by minimizing contact pressure and at the same time maximizing contact area of the cartilage surfaces, our results are consistent with theirs. Both studies have proved that 3D FE model is an efficient tool to predict cartilage contact stress change before and after PAO reorientation planning [8]. However their virtual PAO procedure was roughly performed in Abaqus due to the fact that Abaqus does not have a precise virtual reorientation planning tool and an accurate approach for quantifying 3D hip joint morphology. In addition, their work modeled only a static loading for biomechanics analysis.

In summary, we developed a 3D FE model to predict cartilage contact pressure based on our previously developed computer-assisted planning system for PAO. Our experimental results demonstrated that the developed 3D FE model could be used to find the optimal reorientation of the acetabulum fragment after PAO by minimizing peak contact pressure and at the same time maximizing contact area of the cartilage surfaces. In conclusion, this study suggested that our computer-assisted planning and navigation system with FE modeling can be a promising tool to determine the optimal PAO planning strategy.

Acknowledgement. This project was supported by Strategic Japanese-Swiss Cooperative Program.

References

1. Murphy, S.B., Millis, M.B., Hall, J.E.: Surgical correction of acetabular dysplasia in the adult A Boston experience. Clin. Orthop. Relat. Res. 363, 38–44 (1999)
2. Ganz, R., Klaue, K., Vinh, T., Mast, J.: A new periacetabular osteotomy for the treatment of hip dysplasia. Technique and preliminary results. Clin. Orthop. Relat. Res. 232, 26–36 (1988)
3. Clohisy, J.C., Carlisle, J.C., Trousdale, R., Kim, Y.J., Beaule, P.E., Morgan, P.M., Steger May, K., Schoenecker, P.L., Millis, M.B.: Radiographic evaluation of the hip has limited reliability. Clin. Orthop. Relat. Res. 467(3), 666–675 (2009)
4. Carlisle, J.C., Zebala, L.P., Shia, D.S., Hunt, D., Morgan, P.M., Prather, H., Wright, R.W., Steger May, K., Clohisy, J.C.: Reliability of various observers in determining common radiographic parameters of adult hip structural anatomy. Iowa Orthop. J. 31, 52–58 (2011)
5. Zhao, X., Chosa, E., Totoribe, K.: Effect of periacetabular osteotomy for acetabular dysplasia clarified by three-dimensional finite element analysis. J. Orthop. Sci. 15, 632–640 (2010)
6. Armand, M., Lepisto, J.V.S., Merkle, A.C., Tallroth, K., Liu, X., Taylor, R.H., Wenz, J.: Computer-aided orthopedic surgery with near-real-time biomechanical feedback. Johns Hopkins APL Technical Digest 25, 242–252 (2004)
7. Armiger, R.S., Armand, M., Tallroth, K., Lepist, J., Mears, S.C.: Three-dimensional mechanical evaluation of joint contact pressure in 12 periacetabular osteotomy patients with 10-year follow-up. Acta Orthopaedica 80, 155–161 (2009)
8. Zou, Z., Chavez-Arreola, A., Mandal, P., Board, T.N., Alonso-Rasgado, T.: Optimization of the Position of the Acetabulum in a Ganz Periacetabular Osteotomy by Finite Element Analysis. J. Orthop. Res. 31(3), 472–479 (2013)
9. Liu, L., Ecker, T., Schumann, S., Siebenrock, K., Nolte, L., Zheng, G.: Computer assisted planning and navigation of periacetabular osteotomy (PAO) with range of motion (ROM) optimization. In: Golland, P., Hata, N., Barillot, C., Hornegger, J., Howe, R. (eds.) MICCAI 2014, Part II. LNCS, vol. 8674, pp. 643–650. Springer, Heidelberg (2014)
10. Zheng, G., Marx, A., Langlotz, U., Widmer, K.H., Buttaro, M., Nolte, L.P.: A hybrid CT free navigation system for total hip arthroplasty. Computer Aided Surgery 7, 129–145 (2002)
11. Niknafs, N., Murphy, R.J., Armiger, R.S., Lepist, J., Armand, M.: Biomechanical Factors in Planning of Periacetabular Osteotomy. Biomechanics 1(20), 1–10 (2013)
12. Anderson, A.E., Ellis, B.J., Maas, S.A., Peters, C.L., Weiss, J.A.: Validation of finite element predictions of cartilage contact pressure in the human hip joint. J. Biomech. Eng. 130, 051008 (2008)
13. Caligaris, M., Ateshian, G.A.: Effects of sustained interstitial uid pressurization under migrating contact area, and boundary lubrication by synovial uid, on cartilage friction. Osteoarthritis And Cartilage/OARS. Osteoarthritis Res. Soc. 16, 1220–1227 (2008)
14. Bergmann, G., Deuretzabacher, G., Heller, M., Graichen, F., Rohlmann, A., Strauss, J., Duda, G.N.: Hip contact forces and gait patterns from routine activities. J. Biomech. 34, 859–871 (2001) Including the HIP98 CD, 3-9807848-0-0
15. Phillips, A.T.M., Pankaj, P., Howie, C.R., Usmani, A.S.: Simpso, AHRW: Finite element modelling of the pelvis: Inclusion of muscular and ligamentous boundary conditions. Med. Eng. Phys. 29, 739–748 (2007)

Computer-Assisted Surgical Planning
for Mitral Valve Repair Using 4D Echocardiograms

Mark Hillecke[1], Marco Moscarelli[2], Nilesh Sutaria[2],
Gianni Angelini[2], and Fernando Bello[1]

[1] Simulation and Modelling in Medicine and Surgery, Department of Surgery and Cancer
[2] Imperial College Healthcare NHS Trust
St. Mary's Hospital, Imperial College London, UK
f.bello@imperial.ac.uk

Abstract. The mitral valve is the most commonly diseased valve of the human heart. Depending on the type, nature and severity of the disease, it may be possible to surgically repair the valve. Although tools exist that enable the assessment of the condition of the valve prior to surgery, the details of the surgery plan are typically formulated in the operating theatre. Our work aims to facilitate computer assisted pre-operative planning for mitral valve repair through the provision of a comprehensive set of tools that include visualisation, quantitative methods to assess pathological valves from 4D echocardiograms, and the ability to build a patient-specific, interactive model of the mitral valve that can be used to simulate different types of repair procedures. Initial feedback from subject matter experts indicates that the system has the potential to assist in pre-operative discussion and planning.

1   Introduction

The mitral valve (MV) is the most commonly diseased valve of the human heart. Of the general population, 2–3% suffer from mitral valve prolapse [1], making it the most common cardiac valvular abnormality in industrialised countries [2]. Moreover, Mitral valve stenosis is the most common rheumatic valvular lesion [3].

Transesophageal echocardiography (TEE) has become essential in diagnosing MV disease and gaining insight into the case specific anomalies prior to surgery, with 3D or 4D representations of the MV available in real time. In spite of these advances in imaging, little progress has been made in MV repair preparation beyond efficient visualisation and measurement such as that offered by commercially available software packages like QLAB[1] or 4D MV-Assessment[2].

A recent review of the state-of-the-art of cardiac valve simulation identifies Finite Element Models (FEM) and Fluid Structure Interaction models (FSI) as the most widely used to study valve biomechanics [4]. Lau et al. argue that 'dry' FEM models are probably sufficient if the purpose is to characterise MV closure, without the need for more complex 'wet' FSI models [5]. Burlina et al. proposed an interactive

---

[1] www.healthcare.philips.com
[2] www.tomtec.de

F. Bello and S. Cotin (Eds.): ISBMS 2014, LNCS 8789, pp. 163–172, 2014.

algorithm to build detailed geometrical models of the MV from 3D TEE images and an energy minimization framework to predict the closure of the MV [6]. Later, they extended they work with a number of enhancements in the prediction model and the validation methods [7]. Mansi et al. presented an approach to model MV closure and support the MitralClip procedure, based on patient-specific anatomy obtained from 3D TEE, using a dynamic method, linear tissue models, FEM and tailored boundary conditions [8]. Whilst these models yield clinically relevant insights into valvular morphology and function (e.g. MV closure), they tend to be computationally expensive, unsuitable for real-time clinically oriented simulations [9], and support only limited interactions and repair procedures.

The aim of this work is twofold. First, provision of similar visualisation and measurement tools as in existing commercial applications in order to gain an accurate understanding of the valve's condition and assist in diagnosis. Secondly, the ability to build a real-time interactive model of the MV that may be fitted to patient-specific echocardiographic data and used to simulate different types of repair procedures.

2   Visualisation and Measurement

The first step in any echocardiographic based assessment is always a visual inspection. Volumetric visualisation allows the user to see the entire dataset in 3D and quickly gain a rough idea of the valve's condition. 2D slicing through the 3D dataset is then required for further analysis. We now illustrate and describe the visualisation and quantitative analysis functionality provided by our system.

2.1   Viewports and Scene Graphs

Our mitral valve repair and visualisation system (MiVaRS) provides three independent slice viewports and one volume viewport (Fig. 1). Each viewport maintains its own scene graph to cater for viewport specific differences (e.g. only the volume viewport renders an actual 3D volume). Similarly, each slice viewport is associated with a particular slice plane. A shared scene graph rendered by all viewports is used for shared graphics entities such as the virtual model of the mitral valve.

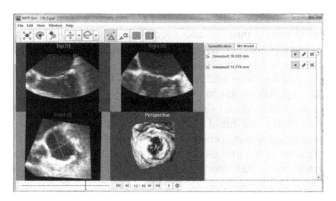

Fig. 1. MiVaRS System – Distance measurements using slice viewports

2.2 Perspective Correction

Commercial scanners such as the Phillips iE33 xMATRIX system[1] store the volumetric echocardiogram data and ECG signal using private DICOM objects that are, for the most part, not publicly documented and/or encrypted. In addition, the raw DICOM data of the echocardiogram requires perspective correction since it is stored as a 'box' shaped volume rather than in its natural frustum shape (Fig. 2a).

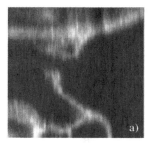 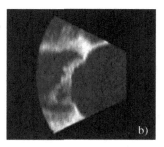

Fig. 2. *a)* Slice rendering from raw DICOM data. *b)* After perspective correction.

Each slice of a 3D echocardiogram consists of a number of rows representing a section of the ultrasound wave echo. Considering every single row as a line segment corresponding to an ultrasound ray, each line will begin at the same distance from the transducer (apex) and have equal length. Our perspective correction consists of two steps that revert the data acquisition and storage process. The initial slice pixel rows (Fig. 3a) are first aligned with the ray sections in terms of their angle to the frustum's apex position in the volume, as determined from the transducer position in the DICOM data. The x axis is assumed to go through the frustum apex and centrally through the bounding box of the volume. The y and z coordinates are then transformed as the x coordinate approaches that of the apex (Fig. 3b). In a second step, the vector from the apex to the n data point on each row is scaled down to the distance from the apex to the n data point in the x direction alone. This ensures the start and end points of each ray match those produced during data acquisition (Fig. 3c). To optimize memory usage and improve rendering quality, perspective correction is performed during rendering rather than data loading (Fig. 2b).

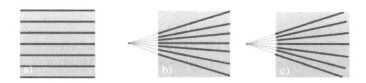

Fig. 3. *a)* Initial slice rows. *b)* Rows aligned with ray sections. *c)* Distance scaling.

2.3   Slice and Volume Rendering

We exploit the features of modern GPUs and GLSL shaders to implement efficient rendering coupled with perspective correction. The entire volumetric image is converted into a volume texture and transferred into GPU memory. Once a slice plane is selected, texture coordinates are calculated by the shader programme, whereas pixel value lookup and interpolation are handled by the GPU.

Volume rendering is done through ray casting implemented as a two pass GLSL shader inspired by [10], with additions and modifications to support single channel echocardiographic data. The behaviour of the ray casting algorithm is influenced by four user controlled factors: quality, alpha threshold, cut-off and roughness (Fig. 4).

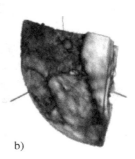

a)   b)

Fig. 4. *a)* Ray casting with no alpha adjustments *b)* Adjusting alpha threshold and cut off

2.4   Measurement

MiVaRS supports four types of measurements specified in a slice viewport: distances, angles, areas and volumes. Distance measurements calculate the distance between two points. Angles measure the angle between two lines. Area measurements indicate both perimeter and resulting area of a user specified polygon. Volumes are computed through a flood fill algorithm with the user specifying a seed point, a threshold and desired number of iterations to control bleeding / leakage. After selecting a measurement tool, the user can perform the necessary actions in the chosen slice viewport, with the other viewports updated accordingly.

3   Virtual Model of the Mitral Valve

A virtual model must be built to simulate the behaviour of a pathological mitral valve. This model should be accurate, yet simple enough to support interactive valve repair simulation. Given these constraints, a mass-spring model (MSM) was chosen. MSM models have been shown to simulate in-plane behaviour considerably faster than FEM, with only minor differences in the deformed mesh [9]. Recognizing the low spatial resolution and comparative low image quality of echocardiograms, and the need to reduce time-consuming manual interactions, we focused on constructing a template model of a generic mitral valve that may be rapidly aligned with a patient-specific 4D echocardiogram, rather than on direct geometric extraction.

3.1   Mass-Spring Model

The mitral valve is composed of anterior and posterior leaflets, mitral annulus and chordae tendineae (Fig. 5a). The leaflets are irregular, membranous flaps of tissues attached to the mitral annulus. The posterior leaflet comprises three scallops (P1-P3). While the anterior leaflet does not have scallops, parts of it have been assigned names corresponding to opposing scallops (A1-A3). There are two types of zones on the leaflets: the coaptation zone at the borders of the leaflets where they connect to the chordae, and the remaining part of the leaflet – the atrial zone. Both leaflets connect to the papillary muscles of the left ventricle by the inelastic chordae tendineae. Our model incorporates leaflets and mitral annulus. Chordae are considered as a constraint, involving an error correction whenever the distance between mass particles connected to the papillary muscles increases beyond the allowed chordae's length.

A polygonal mesh approximating the mitral valve geometry was created to facilitate the construction of the MSM (Fig. 5b). Quadrilaterals were used to simplify modifications to the tissue structure, as well as for reasons of balance and uniformity. The TraerPhysics library 3.0 [11] was used to simulate the movement of the virtual model of the mitral valve.

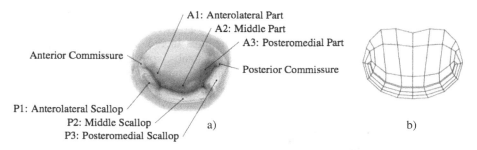

Fig. 5. *a)* Schematic of the Mitral Valve. *b)* Polygonal mesh approximation.

Cross brace springs between the vertices of each quadrilateral are employed to ensure it returns to its original configuration after deformation (Fig 6a, 6b). The length of the cross brace springs must be recomputed dynamically whenever a change to the length of the edge springs is made, for example by a surgical intervention. Bridge springs are added to prevent groups of quadrilaterals from behaving unrealistically, particularly at the union between individual tissue quadrilaterals (Fig. 6c). As with cross brace springs, the length of the bridge springs is calculated dynamically from the lengths of other springs. The original orientation and length of the springs connecting the two mass particles of the bridge spring are taken into account. If their length changes, the length of the bridge spring will be recalculated accordingly.

The material properties of the spring network are preset at the start of a simulation to empirically determined values guaranteeing stability and realistic behaviour.

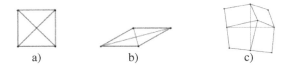

Fig. 6. *a)* Quadrilateral with cross braces. *b)* Cross brace springs are compressed (red) or stretched (blue) as the quadrilateral deforms. c) Bridging springs (red) added to a basic network of mass particles and springs (black).

3.2   Encoding of Model Properties

Texture mapping was used as a simple and efficient way to incorporate the properties of the individual components of the valve into the polygonal mesh. The colour of the pixel to which a vertex maps is used to determine properties of the corresponding mass point. Three textures were employed to encode characteristics of the valve, with a fourth used for visualisation purposes (Fig. 7). Positioning of the vertices has been transformed into a polar coordinate system to simplify texture mapping (Fig. 7a).

The first texture (Fig. 7b) encodes key properties of individual mass points based on a priori knowledge of the MV. Black pixels indicate ordinary mass particles without special properties. Blue pixels correspond to annulus mass points. All particles making up the annular ring will have their positions fixed. Red, green and yellow pixels designate mass points that are connected to the posterior, anterior or both papillary muscles, respectively. The coaptation zone is indicated by the red, green, yellow and grey pixels. Identification of the coaptation zone is crucial to support surgical procedures such as triangular resection. The textures in Fig. 7c and Fig. 7d jointly define a convenient coordinate system to associate a valve coordinate to each point on the mitral valve tissue in terms of distance from the annulus / coaptation zone, and along the direction of the annulus. This coordinate system will be used in some of the algorithms for surgical procedures.

Fig. 7. *a)* Texture coordinates of the mitral valve polygonal mesh vertices. *b)* Texture encoding key properties of individual mass points. *c)* Texture to define the angular coordinate along the direction of the annular ring. *d)* Texture to define the central coordinate towards and away from the annular ring. *e)* Texture used for visual rendering.

3.3   Model Fitting and Animation

The user is provided with tools to manually adjust various parameters in order to fit the virtual model of the valve to the echo data: position, rotation and scaling of the mitral annulus, position of the anterior and posterior papillary muscles. Although this set of parameters might seem limited, our experiments show that it allows a reasonably good fit. Further refinement of the shape is possible in order to capture the effects of mitral valve disease (see 3.4).

4D echo data is typically recorded for the duration of two cardiac cycles (15-25 frames). Aligning all parameters of the virtual model with each frame would be very time consuming. MiVaRS allows users to define key frames between which parameters are linearly interpolated. A reasonably good approximation can be achieved by choosing frames where diastole changes into systole and vice versa. Further frames may be iteratively added as necessary if a higher degree of accuracy is required. Refining the alignment in this manner has the added benefit of only a few parameters having to be adjusted, since the interpolated frame will already closely match the desired alignment, as well as reducing processing time. The resulting animation is stored in MiVaRS.

During play back, the current parameters of the virtual model are updated according to the stored animation. Changes in position, orientation and scale of the valve as a result of the animation will only be applied to mass points making up the annulus of the valve, with all remaining mass points updated as part of the physical simulation.

3.4 Disease Modelling

Two common valve diseases are modelled: chordal ruptures and excess tissue. Chordal ruptures entail the failure of a chord to hold leaflets in place during systole. MiVaRS approximates this by simply removing the chord constraint from the virtual model and updating the physical simulation and graphical representation. Chords are visually represented as lines that may be manually chosen by the user for removal.

Excess tissue is another major cause of mitral prolapse. The user can select the location and amount of excess tissue to be induced into the template shape of the mitral valve. Since excess tissue typically occurs near the coaptation zone, its position can be fully specified by a single angular value using the coordinate system defined above. The amount of excess tissue is defined by specifying the area of tissue affected, as well as the severity of excess tissue. Using the valve coordinate system, the width (angular component) and depth (central component) specify a rectangle in valve space. The effect of the intervention is then modelled by extending the lengths of the springs covered by the selected area according to the severity factor (Fig. 9a).

4 Virtual Mitral Valve Surgery

The modelled valve diseases enable the simulation of two frequent repair procedures: triangular resection and chordal replacement.

4.1 Triangular Resection

The triangular resection shape is defined by a position, width and depth (Fig. 8a). Two cutting planes are automatically generated (green and blue lines) and, together with the angular width of the resection (red line), are used to identify the parts of the

MSS inside the area to be removed. The surgical procedure consists of the resection itself performed by cutting into the tissue from the coaptation zone towards a common point (incision peak point), and sewing the tissue back together (green and blue lines).

To avoid an uneven distribution of springs and mass particles in the network, instead of trying to resect and sew back together tissue (i.e. recreate the procedure itself), we focus on reproducing the outcome by altering the properties of the underlying mass spring network through shortening of spring lengths and reduction of the weight of mass points within the resection area (Fig. 8b). The length of annular springs contained within the resection area is set to zero. Partially contained annular springs are shortened by the amount contained within the resection area. The weight of mass particles contained within the resection area is set to a very small, non-zero value. As a result, mass points with similar central valve coordinate collapse into a single mass point. Supporting springs are adjusted based on the new spring lengths as indicated in section 3.1.

4.2   Chordal Replacement

A new chord is added by choosing a pair of mass particles in the valve and papillary muscle, which in turn determine its length. Papillary muscles are represented by separate mass points to which the new chord may be connected along with already existing chordae. Connecting the chord to the valve entails adding a new mass point and subdividing the mass spring network. All quadrilaterals between the annular ring and coaptation zone at the angular level of the new mass point are split in two (Fig. 8c). After subdivision of the tissue, the chord can be created at the required length with connections to the user specified papillary muscle and the newly created mass point at the tip of the leaflet (Fig. 9b). Supporting springs are adjusted according to the new configuration as indicated in section 3.1.

Fig. 8. *a)* Triangular resection & key points. *b)* Planned resection (red – fully contained; blue – partially contained). c) Chord replacement subdivision (blue – new mass point).

5   System Evaluation

5.1   Expert Questionnaire

MiVaRS was demonstrated to two specialists in mitral valve disease (Cardiothoracic Surgeon and Cardiologist). After using the system, they completed a questionnaire regarding its quality and usefulness. The overall impression of the system and ease of use was positive (M=4.83 S=0.40) on a 5 point Likert scale. Slice and volume visualisation (controls, quality, ease of navigation), measurement tools (type, usefulness, ease of use) and virtual model and simulation (realism, configuration options, physical model, frame animation) also rated highly (M=4.75 S=0.46; M=4.83 S=0.40, M=4.80 S=0.42 respectively). Surgical interventions (configuration and ease of use) rated the highest with unanimous agreement (M=5 S=0). Improvements suggested include: slicing on volume viewport, Simpson's biplane method to define volumes, modelling coaptation surface, configurable number / location of chordae, slicing of virtual model, implanting annuloplasty rings, simulating flailing chordae, Doppler echocardiography and virtual regurgitant jet.

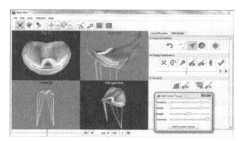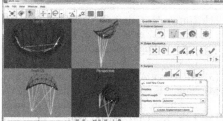

Fig. 9. *a)* Excess tissue added to the anterior leaflet. *b)* Chordal replacement (white line).

5.2 Visualisation and Measurement

The correctness of the visualisation functionality (4D echo data load, perspective correction, slice rendering) and accuracy of the measurements were validated on a sample dataset through a direct comparison between results obtained using MiVaRS and those from the Phillips iE33 xMATRIX built-in software. Table 1 shows a mean error of less than 0.5mm, likely caused by user inaccuracies during points placement.

Table 1. Quantification error for common distance measurements

Measurement	iE33	MiVaRS	Δ [mm]	Δ [%]
Anteroposterior Dimension	30.6 mm	30.02 mm	0.58 mm	1.89 %
Lateral Dimension	32.1 mm	31.78 mm	0.32 mm	0.99 %
Anterior Leaflet Length	29.7 mm	29.34 mm	0.36 mm	1.21 %
Average			0.42 mm	1.37 %

6 Conclusions and Future Work

A prototype software package (MiVaRS) providing a comprehensive solution for visualisation and measurement of the mitral valve, together with real-time interactive simulation for mitral valve repair using 4D echocardiography data has been designed, implemented and evaluated by two specialists in mitral valve disease. Surgical and medical experts confirmed the usefulness and quality of MiVaRS, including the resemblance of the virtual mitral valve to the real valve, whilst quantitative evaluation indicated that its visualisation functionality and measurement accuracy is comparable to existing proprietary software.

Contemplated future work includes quantitative analysis of the virtual mitral valve surgery by comparing the resulting simulation with post-operative 4D echocardiograms. We will also investigate incorporating suggestions from the experts (see 5.1) and different strategies for determining the material properties of the MSS.

References

[1] Hayek, E., Gring, C.N., Griffin, B.P.: Mitral valve prolapse 365, 507–518 (2005)
[2] Freed, L., et al.: Mitral valve prolapse in the general population: the benign nature of echocardiog. features in the framingham heart study. JAC of Card. 40(7), 1298–1304 (2002)
[3] Mebazaa, A., et al.: Acute Heart Failure. Springer (2007)
[4] Votta, E., et al.: Toward patient-specific simulations of cardiac valves: State-of-the-art and future directions. Journal of Biomechanics 46, 217–228 (2012)
[5] Lau, K., Diaz, V., Scambler, P., et al.: Mitral valve dynamics in structural and fluid-structure interaction models. Medical Eng. & Phys. 329, 1057–1064 (2010)
[6] Burlina, P., Sprouse, C., DeMenthon, et al.: Patient-specific modeling and analysis of the mitral valve using 3D-TEE. In: IPCAI, pp. 135–146 (2010)
[7] Burlina, P., Sprouse, C., Mukherjee, R., et al.: Patient-Specific MV Closure Prediction using 3D Echocardiography. Ultrasound Med. Biol. 39(5), 769–783 (2013)
[8] Hammer, P., Sacks, M., del Nido, P., et al.: Mass-Spring Model for Simulation of Heart Valve Tissue Mechanical Behavior. Ann. Biom. Eng. 39(6), 1668–1679 (2011)
[9] Kruger, J., Westermann, R.: Acceleration techniques for gpu-based volume rendering. In: Proc. 14th IEEE Visualization 2003 (VIS 2003), p. 38. IEEE Computer Society (2003)
[10] Bernstein, J.T.: TraerPhysics Library 3.0. (2012), http://murderandcreate.com/physics/ (last accessed August 14, 2014)

Generic 3D Geometrical and Mechanical Modeling of the Skin/Subcutaneous Complex by a Procedural Hybrid Method

Christian Herlin[1,2,3], Benjamin Gilles[4], Gérard Subsol[4],
and Guillaume Captier[1,3]

[1] Dept. of Plastic Pediatric Surgery, CHRU Montpellier, France
c-herlin@chu-montpellier.fr
[2] Dept. of Plastic Surgery, Burns and Wound Healing, CHRU Montpellier, France
[3] Laboratory of Anatomy, Montpellier 1 University, France
[4] ICAR Research Team, LIRMM, CNRS/University of Montpellier 2, France

Abstract. The aim of this work is to build a 3D geometric and mechanical model of the skin/subcutaneous complex (SSC) which could be adapted to the different parts of the body and to the morphological parameters of the patient. We present first the anatomical pattern of the SSC. Then, we propose a hybrid model which combines volume, membranous and unidimensional models. The complex internal structure of the SSC is automatically created by a procedural process. All the models are defined by some parameters which can be easily measured by medical imaging. We describe several preliminary experiments which show how this hybrid method models realistic geometrical deformations and physical behaviors and could be used for surgery simulation and planning.

Keywords: biomechanics, soft tissue, skin/subcutaneous complex, plastic surgery, procedural method.

1 Introduction

Surgeons, biomechanicians and digital designers are all looking for accurate, realistic means of predicting the deformation of soft tissues. Although different objectives have been pursued in each case, the main problem arising has been that of modelling a heterogeneous three-dimensional structure showing complex mechanical behavior. In fact, there exists no mechanical model so far which accurately simulates the anatomy of the skin/subcutaneous complex (SSC), taking the connective arrangements and the mobility of the components into account.

Most models are uni- or bi-lamellar, composed of layers which often have a constant thickness and which show linear isotropic mechanical behavior (see for example [5,4,9,14,10] for the face or the breast). They differ considerably from the genuine architecture of the human SSC, which is supported by a complex three dimensional collagen network causing a non-linear anisotropic mechanical behavior. Moreover, these models are dedicated to a specific anatomical area.

F. Bello and S. Cotin (Eds.): ISBMS 2014, LNCS 8789, pp. 173–181, 2014.
© Springer International Publishing Switzerland 2014

Although the structure of subcutaneous tissue differs between various regions of the human body, common components have been identified in several anatomical parts. The existence of a generic pattern of organization of the SSC was put forward several years ago [11], then confirmed in recent radiological work [1] at the level of the arms, legs and trunk. Nevertheless, this anatomical finding has never been the basis of a generic modeling of the human SSC.

The aim of this work is to build, from anatomical knowledge, medical imaging and histological observation, a generic 3D geometric and mechanical model of the skin/subcutaneous complex which could be adapted to the different parts of the body and to the morphological parameters of the patient. In section 2, we present the anatomical pattern of the SSC. In section 3, we propose a hybrid model which combines volume, membranous and unidimensional models of the different substructures. The complex internal structure of the SSC is automatically created by a procedural process. All the models are defined by some parameters which can be easily inferred by medical imaging. In section 4, we describe several preliminary experiments which show how this method models realistic geometric deformations and physical behaviors and could be used for surgery simulation and planning.

2   Anatomical Considerations

2.1   Generic Pattern of the Skin/Subcutaneous Complex

The generic model of organization of the human skin/subcutaneous complex was confirmed by our 3T MRI studies at the level of the face, the trunk and the limbs [7]. We can summarize this organization into a multi-layered model retained by a collagen network which combines the following substructures (see Fig. 1):

- **Skin** which has a very variable thickness (1 to 7 mm) depending on locations;
- **Superficial Adipose Tissue** (SAT) consisting of fat lobules of variable size organized in clusters rather like palisades bands;
- **Retinacula Cutis Superficialis** (RCS) formed by the septal densifications lobules of the SAT;
- **Membranosum Stratum** (SM) or membranous layer of the superficial fascia; it can be either fibrous and/or muscular and either single or multiple;
- **Deep Adipose Tissue** (DAT) composed of polyhedral fat lobules, as in a honeycomb arranged more randomly than in the SAT;
- **Retinacula Cutis Profondus** (RCP) formed by the septal densifications lobules of the DAT;
- **Skin ligaments** (SL) [13], also perpendicular to the skin which are a particular specialization of RCS and RCP . They allow the maintenance of the skin or SM in deep plane or skin. They have already been described at the face, breast, the fingers, the sole or the scalp.
- **Deep Layer** which is the inferior limit of the SSC. This may be muscle, aponeursosis or periosteum.

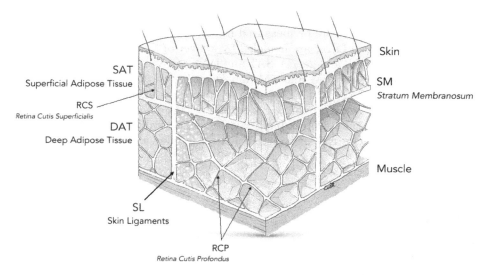

Fig. 1. Generic model of the Skin/Subcutaneous Complex with all the substructures

2.2   Regional Specificity

At the level of the face, the SM endorses a particular muscle specialization called *Superficial Musculo Aponevrotic System* (SMAS) ensuring facial expressions [9]. This residue of the panniculus carnosus is found to be in continuity with the temporoparietal fascia and the superficial layer of the cervical fascia. However, this continuity is interrupted between the two nasojugal fold where the layers are intermingled, making the stratigraphic pattern more difficulty to identify.

At the level of the breast and thorax, the SM identified beyond the mammary region is continuous anteriorly, in the breast area, with a superficial layer located 3 to 4 mm under the skin. The collagen network, less developed in male chest, corresponds to the Cooper's ligament in female breast.

In the antero-lateral part of the abdomen, the SM forms a single layer. More anteriorly, it is divided into two parts when the adipose tissue becomes thicker. In the peri-umbilical region, the SM takes a deeper course and merges with the pre-muscular fasciae of the rectus abdominis muscles.

Finally, for the lower limbs and buttocks, the SM and Retinacula Cutis show a similar pattern of organization. A clear-cut increasing gradient in the number of RC in the knee and buttocks, correspond to the preferential fat storage sites.

3   3D Modeling of the Skin/Subcutaneous Complex

3.1   Procedural Geometrical Modeling

In order to deal with the variability of the SSC with respect to the regional specificity or the morphology of the patient, we introduce a **procedural** method to build a detailed 3D geometrical model. This model is based on some global parameters which can easily be measured in MRI or echography.

Skin, SM and Deep Layer are modeled by a layer of a given thickness (called respectively $skin_t$, SM_t and $deep_t$). SAT and DAT correspond respectively to the spaces between skin and SM and SM and Deep Layer, and are parameterized by their thickness SAT_t and DAT_t. These spaces are filled by a 3D Voronoï tessellation where each Voronoï cell, defined by its center point, represents a fat lobule. Center points are randomly located but evenly-spaced by using the Lloyd's algorithm. The mean diameter of a fat lobule called $lobule_d$ gives the average distance between the center points.

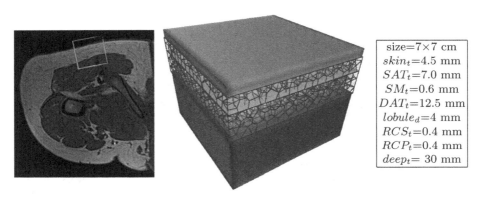

| size=7×7 cm |
| $skin_t$=4.5 mm |
| SAT_t=7.0 mm |
| SM_t=0.6 mm |
| DAT_t=12.5 mm |
| $lobule_d$=4 mm |
| RCS_t=0.4 mm |
| RCP_t=0.4 mm |
| $deep_t$= 30 mm |

Fig. 2. SSC model for the buttocks based on parameters measured in a MR image

The walls of the Voronoï tessellation correspond to the septa of RCS and RCP which separates the lobules. They are extracted as a 3D triangle mesh. We assume plane stress conditions for these walls, meaning that we neglect bending over stretching; so wall thickness is not geometrically represented.

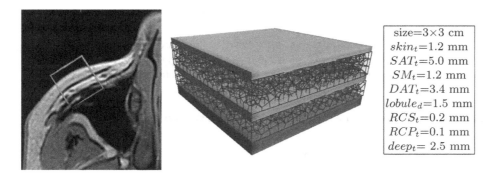

| size=3×3 cm |
| $skin_t$=1.2 mm |
| SAT_t=5.0 mm |
| SM_t=1.2 mm |
| DAT_t=3.4 mm |
| $lobule_d$=1.5 mm |
| RCS_t=0.2 mm |
| RCP_t=0.1 mm |
| $deep_t$= 2.5 mm |

Fig. 3. SSC model for the cheek based on parameters measured in a MR image

Fig. 2 and 3 demonstrate that specificity of the anatomical region is taken into account. At the level of the buttocks, the skin is thick, the thickness of the DAT is very variable depending on the location. The mesh formed by the RCS and RCP is very dense in areas of fat storage. At the level of the face, the SM, thicker, has a muscle differentiation. The volume of lobules is small and RCS as RCP are thin.

3.2   Hybrid Mechanical Modeling

From a mechanical point of view, the SSC is composed of elements of variable mechanical behaviors: hyperelastic for the skin and visco-elastic for the fat. This assembly has a global non-linear anisotropic behavior principally due to the existence and orientation of collagen fibers. Detailing the stratigraphic and lobular organization of the sub-cutaneous fat, we wanted to recreate the overall anisotropy of the SSC by adding linear isotropic elements. This approximation is permitted by the architectural complexity of the geometrical model.

To model the different substructures, we built an **hybrid** model. It combines a volume model for the skin, SM, SAT and DAT, a membranous model for the interlobular septa of RCS and RCP and a unidimensional model for the skin ligaments.

We used the SOFA framework [3] which is particularly suitable for implementing and processing this type of composite models in nearly real-time. Programming is based on a Python script that allows flexibility in the geometrical and mechanical parameterization.

Volume, membranous and unidimensional components of our model were discretized using respectively hexahedral, triangular and edge finite elements. To mechanically couple them, these elements were embedded in a coarser regular grid using barycentric interpolation. The typical size of a coarse hexahedron was 2 mm. Time integration for coarse nodes was performed using the Euler implicit scheme and conjugate gradient algorithm. Strain was measured using the corotational method.

For each substructure, we extracted Young's modulus data from the literature, and performed an average. We try to assess these values by a simulation of indentation in similar conditions of some published research. Poissons ratio was fixed at 0.49 due to the incompressible nature of the materials.

4   Some Experiments

4.1   Simulation of Indentation

In order to validate the behavior of our model, we simulated an indentation test of the forearm with the parameters found in an in vivo MRI study [15]. Geometrical parameters are listed in Fig. 4, right. In paticular, DAT is very thin. Young modulus of skin, fat, SM, RC and muscle were respectively set to 20, 1 , 60, 100 and 15 kPa.

The indenter was modeled as a cylinder of diameter 5 mm. Such as in the experiment presented in [15], we simulated several indentations between 0.35 and 1.8 N. Fig. 4, shows a comparison between real data and the obtained vertical displacements at equilibrium. On average the error was 18% suggesting that our model is rather realistic.

Accuracy could be improved in future using non-linear material laws. However, this will raise the problem of parameter identification, as a high discrepancy in the reported material parameter can be found in the literature, depending on the measurement modality and test specimen.

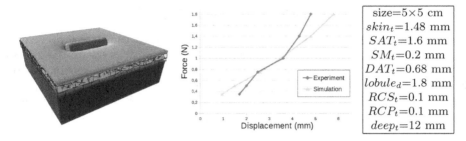

size=5×5 cm
$skin_t=1.48$ mm
$SAT_t=1.6$ mm
$SM_t=0.2$ mm
$DAT_t=0.68$ mm
$lobule_d=1.8$ mm
$RCS_t=0.1$ mm
$RCP_t=0.1$ mm
$deep_t=12$ mm

Fig. 4. Simulation of indentation of the forearm according to an in vivo MRI study

4.2   Simulation of the Phenomenon of Cellulite

Cellulite is considered for the majority of authors as a consequence of the increased content of fat lobes associated with a deterioration of collagen tissue of the skin and the relative stability of RC underlying the skin [8]. We simulated the increase of 24 cm^3 of fat in the SAT and DAT of the buttocks model of Figure 2 by modifying the rest shape of fat material. This results in skin dimpling and nodularity as shown in Fig. 5 left. Then we reduced the stiffness of the skin to simulate aging. Cellulite is enhanced as we can see in Fig. 5 right.

4.3   Simulation of Fat Injection in the Subcutaneous Tissue

Fat grafting is a major technique in reconstructive surgery. During this procedure, the injection site significantly affects the outcome. In a deep injection (that is in DAT), the effect is said "volumizer" and the main purpose is usually to increase the projection of an anatomical structure (cheekbone, buttocks, breasts...). In contrast, if the injection is more superficial, the effect is more on the shape of the injection site and it is most often used to fill wrinkles or scars.

We simulated an injection of 0.5 cm^3 of fat inside the DAT of the cheek model of Fig. 3. It results in a smooth and global elevation of the skin surface (see Fig. 6 left). Fig. 6 right shows the simulation of a similar injection inside the SAT. Our model allows to simulate the "orange peel" phenomenon with a high fidelity. This problem during surface injection is common in daily clinical practice and feared if a too large volume of injection is carried out in a too superficial plane.

Fig. 5. Simulating cellulite in buttocks by increasing fat volume in SAT and DAT. Right: cellulite enhancement due to aging simulation.

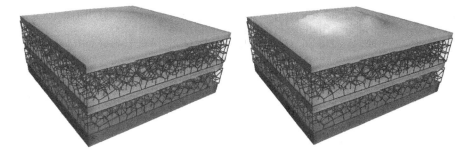

Fig. 6. Simulating fat injection in the cheek at the level of DAT (left) and SAT (right). If the injection is too superficial, the "peel orange" phenomenon appears.

5   Future Work

If we have the external shape of the patient, for example by using MRI or a 3D surface scanner, we could deform the generic model and fit it to his specific and exact morphology. Such a registration will allow us to perform pre-operative simulations of plastic surgeries of the face, the thorax or the breast. In particular, we think that our model will be able to simulate breast augmentation surgery in a more faithful way because it includes an accurate modeling of RC, which corresponds to the Cooper's ligament which has a major role in the final shape of the breast.

This model will also permit to simulate fat grafting inside the adipose tissue taking into account its lobular composition and its retaining connective tissue. Some studies focus on fat injection as [6] and more recently [12] but none is able to simulate fat grafting considering the destruction of septa or skin ligaments. Such a technique called fasciotomy is common in daily practice, in a particular in breast reconstruction [2]. The simulation could be done procedurally by deleting the walls of the Voronoî cells corresponding to the RCS and cutting the skin ligaments.

Nevertheless, to simulate a more accurate behavior of the skin, it will be necessary to add to our model the anisotropic behavior of the skin and the pre-existing tension. These properties have been described in details and are often represented as Langer's lines. This would allow us to simulate other type of plastic surgeries as skin incision or lifting.

Acknowledgments. We want to thank the radiology team of University Hospital of Nîmes, France for MR acquisitions and Ms. Witt for the quality of her drawings.

References

1. Abu-Hijleh, M.F., Roshier, A.L., Al-Shboul, Q., Dharap, A.S., Harris, P.F.: The membranous layer of superficial fascia: evidence for its widespread distribution in the body. Surgical and Radiologic Anatomy 28(6), 606–619 (2006)
2. Agha, R., Fowler, A., Herlin, C., Goodacre, T., Orgill, D.: Use of autologous fat grafting for reconstruction post-mastectomy and breast conserving surgery: A systematic review and meta-analysis. European Journal of Surgical Oncology 40(5), 614–615 (2014)
3. Allard, J., Cotin, S., Faure, F., Bensoussan, P.J., Poyer, F., Duriez, C., Delingette, H., Grisoni, L.: SOFA - an open source framework for medical simulation. In: MMVR 15 - Medicine Meets Virtual Reality, vol. 125, pp. 13–18 (February 2007), http://hal.inria.fr/inria-00319416
4. Barbarino, G., Jabareen, M., Trzewik, J., Mazza, E.: Physically based finite element model of the face. In: Bello, F., Edwards, E. (eds.) ISBMS 2008. LNCS, vol. 5104, pp. 1–10. Springer, Heidelberg (2008)
5. Chabanas, M., Luboz, V., Payan, Y.: Patient specific finite element model of the face soft tissues for computer-assisted maxillofacial surgery. Medical Image Analysis 7(2), 131–151 (2003)
6. Comley, K., Fleck, N.: Deep penetration and liquid injection into adipose tissue. Journal of Mechanics of Materials and Structures 6(1-4), 127–140 (2011)
7. Herlin, C., Chica-Rosa, A., Subsol, G., Gilles, B., Macri, F., Beregi, J., Captier, G.: Three-dimensional study of the skin/subcutaneous complex using in vivo whole body 3 tesla MRI. review of the literature and confirmation of a generic pattern of organisation. Accepted for Publication in Surgical and Radiologic Anatomy (2014)
8. Hexsel, D., Siega, C., Schilling-Souza, J., Porto, M.D., Rodrigues, T.C.: A comparative study of the anatomy of adipose tissue in areas with and without raised lesions of cellulite using magnetic resonance imaging. Dermatologic Surgery 39(12), 1877–1886 (2013)
9. Hung, A.P.L., Wu, T., Hunter, P., Mithraratne, K.: A framework for generating anatomically detailed subject-specific human facial models for biomechanical simulations. The Visual Computer, 1–13 (May 2014)
10. Lapuebla-Ferri, A., del Palomar, A.P., Herrero, J., Jiménez-Mochol, A.J.: A patient-specific FE-based methodology to simulate prosthesis insertion during an augmentation mammoplasty. Medical Engineering & Physics 33(9), 1094–1102 (2011)
11. Lockwood, T.E.: Superficial fascial system (SFS) of the trunk and extremities: a new concept. Plastic and Reconstructive Surgery 87(6), 1009–1018 (1991)

12. Majorczyk, V., Cotin, S., Duriez, C., Allard, J.: Simulation of lipofilling reconstructive surgery using coupled eulerian fluid and deformable solid models. In: Mori, K., Sakuma, I., Sato, Y., Barillot, C., Navab, N. (eds.) MICCAI 2013, Part III. LNCS, vol. 8151, pp. 299–306. Springer, Heidelberg (2013)
13. Nash, L.G., Phillips, M.N., Nicholson, H., Barnett, R., Zhang, M.: Skin ligaments: regional distribution and variation in morphology. Clinical Anatomy 17(4), 287–293 (2004)
14. Roose, L., De Maerteleire, W., Mollemans, W., Maes, F., Suetens, P.: Simulation of soft-tissue deformations for breast augmentation planning. In: Harders, M., Székely, G. (eds.) ISBMS 2006. LNCS, vol. 4072, pp. 197–205. Springer, Heidelberg (2006)
15. Tran, H.V.: Caractérisation des propriétés mécaniques de la peau humaine in vivo via l'IRM. Ph.D. thesis, Université de Technologie de Compiègne (October 2007)

The MAP Client: User-Friendly Musculoskeletal Modelling Workflows

Ju Zhang[1], Hugh Sorby[1], John Clement[2], C David L Thomas[2], Peter Hunter[1], Poul Nielsen[1], David Lloyd[3], Mark Taylor[4], and Thor Besier[1]

[1] Auckland Bioengineering Institute, University of Auckland, Auckland, New Zealand
{ju.zhang,t.besier}@auckland.ac.nz
[2] Melbourne Dental School, University of Melbourne, Melbourne, Australia
[3] Griffith Health Institute, Griffith University, Queensland, Australia
[4] School of Computer Science, Engineering and Mathematics, Flinders University, Adelaide, Australia

Abstract. Subject-specific models of the musculoskeletal system are capable of accurately estimating function and loads and show promise for clinical use. However, creating subject-specific models is time-consuming and requires high levels of expertise. To address these issues, we have developed the open source Musculoskeletal Atlas Project (MAP) Client software. The MAP Client provides a user-friendly interface for creating musculoskeletal modelling workflows using community-created plug-ins. In this paper, we discuss the design of the MAP Client, its plug-in architecture and its integration with the Physiome Model Repository. We demonstrate the use of MAP Client with a subject-specific femur modelling workflow using a set of modular open source plug-ins for image segmentation, landmark prediction, model registration and customisation. Our long-term goal is to foster a community of MAP users and plug-in developers to accelerate the clinical use of computational models.

Keywords: musculoskeletal modelling, biomechanical modelling, Musculoskeletal Atlas Project, personalised simulation, open source software, pipelines, workflows.

1 Introduction

Computational models play a vital role in understanding structural form-function relationships and mechanisms of injury and disease of the musculoskeletal system. This knowledge is also critical in designing, testing, and validating orthopaedic devices and implants. However, the predictive power of computational models is dependent on the ability to accurately capture the complex geometry of the musculoskeletal structures[3]. Musculoskeletal model generation is a wide and highly active field of research. Topics range from image analysis to computational geometry to statistical modelling. Implementing and using methods of such breadth and depth required a high level of expertise not to mention investment of time. In other fields, a number of projects have been successful in

F. Bello and S. Cotin (Eds.): ISBMS 2014, LNCS 8789, pp. 182–192, 2014.

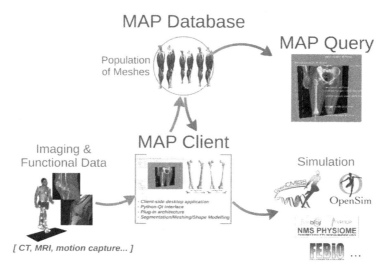

Fig. 1. The MAP Framework. The MAP Client imports images and functional data and facilitates segmentation and meshing using the MAP Database population. A MAP Query tool can determine anatomical features across the population. Meshes exported from the MAP Client are compatible with various simulation environments.

consolidating computational methods into standardised software libraries, e.g. InsightToolkit[12] and CGAL[1]. At a higher workflow-based level, GIMIAS[9] focuses on cardiac modelling workflows, while MedINRIA[2], MeVisLab[3], and Slicer3D[4] provide medical image visualisation and processing frameworks.

The Musculoskeletal Atlas Project (MAP) aims to establish a unified database of musculoskeletal models as well as computational methods for research and clinical application in the field of orthopaedic biomechanics. Three software components make up the MAP (Fig.1): the MAP Database, an online repository for modelling data and workflows; the MAP Query Tool, for extracting population-based metrics from the MAP Database; and the MAP Client, a plug-ins based workflow software for model generation with access to the MAP Database.

This paper focuses on the open source MAP Client which enables researchers to rapidly generate accurate models of the musculoskeletal system from imaging and other data, using a toolbox of community-developed plug-ins. Such a collection of tools in a common software framework bridges an important gap in musculoskeletal modelling. Patient-specific data can be collected from a variety of sources (e.g. CT, MRI, fluoroscopy, motion-capture) and analysed or simulated using a number of established open source and proprietary biomechanical

---

[1] Computational Geometry Algorithms Library, http://www.cgal.org
[2] http://med.inria.fr
[3] http://www.mevislab.de
[4] http://www.slice.org

simulation software (e.g. OpenSim [4], ABAQUS[5]). However, there does not exist an open source toolbox for processing the patient specific data into digital models for the simulation software. Model generation often involves a mishmash of different software, many of which are close sourced, unpublished, or specific to a particular project. These reasons make them hard to share with other researchers, hard to validate, and results in duplication of effort in reimplementing the same algorithms across many projects. When software expertise is not available, algorithms can be implemented badly or not implemented at all, jeopardising the quality of research.

The MAP Client aims to consolidate the model generation methods of the musculoskeletal modelling community not by recreating the multitude of different methods but by allowing groups to wrap their software into MAP Client plug-ins so that they can be shared with and used in a common software framework. The consolidation extends beyond individual methods by being able to save and share entire model-generating workflows composed of a sequence of configured plug-ins. Another novel aspect of the MAP Client is its integration with the MAP Database, which allows users to access a library of models generated and uploaded by other MAP users. The benefits of this database are that it encourages the use of standard, validated models; the origins and histories of models are recorded (e.g. the MAP Client workflow used to generate it) so that it can be recreated or validated; and datasets can be easily shared and worked on using a common set of tools (MAP Client plug-ins).

In the next chapter, we outline the design of the MAP Client, explaining its handling of plug-ins and workflows, and integration with the MAP Database. Section 3 illustrates the use of the MAP Client in a real-world problem: creating patient-specific femur models from MRI and motion-capture (mo-cap) data. We explain the construction of the workflow and provide details on the key plug-ins used. Finally, the key features of the MAP Client are summarised in section 4.

2   The MAP Client Software

The MAP Client is a cross-platform desktop application written in Python based on the Qt widget library[6]. The application is fundamentally a framework for managing workflows. A MAP Client *workflow* is a collection of *steps* connected together to define a particular process. Each step is implemented in software as a plug-in. Therefore, a MAP Client plug-in will be referred to as a step. The MAP Client is able to describe any process using appropriate steps, but is focused on providing those useful for processes used in the musculoskeletal and bioengineering communities.

The design of the MAP Client is set around two main objectives: ease-of-use and community engagement. These two objectives were taken into account when determining which technologies to base the MAP Client on. To meet these objectives the MAP Client is written in the Python language which has been widely

---

[5] Dassault Systemes, Vlizy-Villacoublay, France.
[6] http://qt-project.org/

adopted by the scientific community, and employs a simple plug-in architecture (see below).

The MAP Client is currently released under the GNU General Public Licence version 3 and it is available for download from `https://launchpad.net/mapcl ient` with documentation available at `http://map-client.readthedocs.org/ en/latest/`.

2.1 Plug-In Architecture

Plug-ins lie at the heart of the MAP framework by providing workflow steps. The plug-in framework is lightweight, requires no external libraries, and compliant with both Python 2 and Python 3. These design choices allow steps to be implemented as simply as possible to encourage community involvement. The plug-in interface is defined in documentation with the expectation of compliance placed on developers.

In short, a step is implemented as a Python class with a number of expected class methods called upon by the plug-in interface. Class methods `setPortData` and `getPortData` are required for the interface to input data to the step via input ports, and get data from the step via output ports, respectively. Input and output ports, and the data expect at each port, are defined in the class `__init__` method. Input and output ports of the same data type are allowed to be connected in a workflow. Class method `execute` is required for the interface to run the step (process input data/produce output data), and should execute the main algorithm of the step. Other class methods can be optionally implemented to allow step configuration or the serialisation of step state and configurations.

The Plug-in Wizard incorporated in the MAP Client guides the developer in defining a step and generates a skeleton step class that conforms to the plug-in interface. It is then up to the developer to implement the necessary class methods.

We acknowledge that there may be resistance in spending resources modifying existing software for the MAP Client, which is why the plug-in framework imposes minimal requirements in terms of required libraries or data structures. Any Python or Python-bound libraries can be used by the step, including Qt-compatible libraries for visualisation (e.g. VTK, Matplotlib). Of course, there is no requirement for a plug-in to be completely implemented in Python. For example, the plug-in can simply call on Python-bound C libraries or even simpler, run pre-compiled executables. The choice of using the Python language enables rapid development of plug-ins in a high-level scripting language, cross-platform support, and access to a wide range of open source libraries. A selection of published steps can be found at `https://github.com/mapclient-plugins`.

2.2 MAP Database Integration

The MAP Database is built on the Physiome Model Repository (PMR) [13] and stores musculoskeletal models, images, experimental data, and MAP Client workflows. PMR possesses a version control system so that changes to resources are recorded and their history can be inspected. The PMR Tool in the MAP Client connects the application to the MAP Database via PMR webservices. Users can use this tool to search for suitable resources and download local copies. The PMR Tool can also be accessed through the plug-in interface, so that workflow steps can retrieve or store data within the MAP Database.

Having access to existing models, modelling data, and workflows is a point of difference between the MAP Client and other workflow management software. Some, such as Slicer3D, have online repositories but only for plug-ins. However, resources such as atlas models, statistical models, and training data are crucial for the workflows envisioned for the MAP Client. Through the MAP Database, users can access a library of models created by other MAP users as well as the tools used in their creation. Not only does this encourage collaboration and data-sharing, it also means that modelling resources can be portable, backed-up, and version-controlled.

2.3   User Interface

The MAP Client GUI is composed of three areas: the *Menu Bar*, the *Step Box*, and the *Workflow Canvas* (Fig.2). The Menu Bar provides a number of drop-down menus for accessing the applications functions such as saving/loading workflows, undo/redo, the PMR Tool, step management tools, and accessing documentation. The Step Box shows the selection of steps that are currently installed. The Workflow Canvas is where workflows are constructed and edited, and where steps can be configured.

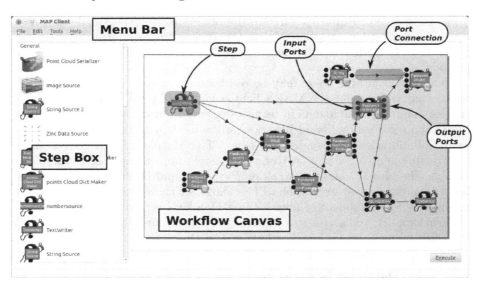

Fig. 2. MAP Client GUI. Installed steps are listed in the *Step Box*. These can be dragged and dropped onto the *Workflow Canvas* to assemble workflows. steps are connected by clicking and dragging between their input and output ports.

2.4   Workflows

There are two types of workflows that can be created: local workflows and PMR workflows. A local workflow will create a project on the host computer while a PMR workflow will create a project on the MAP Database and make it available to the host computer by creating a local copy. PMR workflows can be downloaded by other MAP Database users.

Workflows are created by dragging and dropping steps from the Step Box onto the Workflow Canvas. Dragging between step ports then creates connections to define the complete workflow. Saving a workflow records the arrangement of its steps and their states. Once the workflow is saved, the workflow is executed by clicking the *Execute* button. The intuitive user interface enables complex workflows to be constructed and executed by non-programmer users.

3   Example Workflow: Patient-Specific Femur Modelling

To demonstrate the use of the MAP Client for a real-world problem, we present a MAP Client workflow for generating femur models. Creating patient-specific femur models to predict *in vivo* knee joint loading is a common task in musculoskeletal modelling[7]. Femur position and general size can be reconstructed from sparse anatomical landmarks in mo-cap data. Detailed distal femur geometry can be segmented from high resolution knee magnetic resonance images (MRI). By using a statistical shape model of the femur, an atlas femur mesh can be customised to the landmarks and segmentation to reconstruct an accurate patient-specific femur model. A MAP Client workflow has been created for the tasks above and is presented here with associated key steps (Fig.3).

The workflow begins by the TRC Source step reading and outputting a set of landmarks from mo-cap data in the TRC format. The Hip Joint Centre (HJC) Prediction step then estimates the positions of the centres of the femoral heads based on the other landmarks present. The landmarks are then passed to the Landmark Registration step where an atlas femur mesh is morphed to the femoral landmarks, according to a femur statistical shape model. On a parallel path, the Image Source step loads an MRI stack and sends it to the Manual Segmentation step for segmentation. The Point Cloud Registration step aligns the segmented point cloud to the femur mesh before they are passed to the Principal Components Fitting step which fits the femur mesh to the point cloud according to the femur shape model. Finally, the Parametric Mesh Fitting step refines the fit by adjusting distal femur mesh control points without the constraint of the shape model.

The mesh output from the workflow is of sufficient accuracy (0.3 to 4.3 mm RMSE) for patient-specific modelling of knee kinematics and kinetics. Validation results of each individual step are presented below. The mesh can be output to a variety of common formats using appropriate output steps. The run time of the entire workflow excluding manual segmentation is approximately three minutes. The key steps above are explained below. Of course, these steps can be assembled in many other ways to accomplish other tasks.

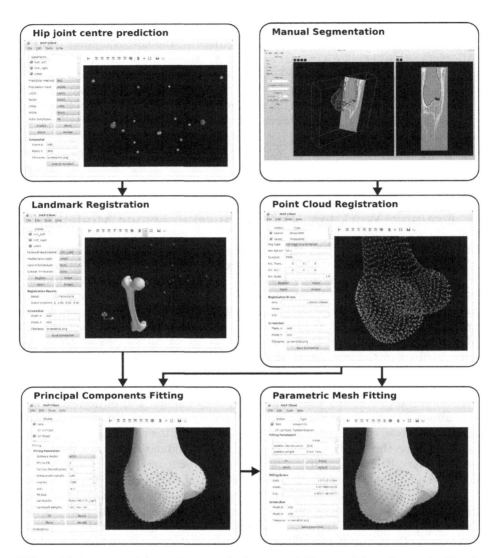

Fig. 3. Major steps of the patient-specfic femur modelling workflow. Anatomical land-marks from mo-cap data and the Hip Joint Centre Prediction step are used to register a femur atlas mesh in the Landmark Registration step. Point clouds from the Manual Segmentation step are registered to the femur mesh in the Point Cloud Registration step. The registered point clouds and atlas mesh are passed to the Principal Components Fitting step and the Parametric Mesh Fitting step where the mesh is fitted to the segmented data. The end product is a patient-specfic femur reconstructed from mo-cap data and knee MRI.

Manual Segmentation Step. This is a tool to manually create a point cloud from a stack of images. Step input is a Python object that describes the image stack, location on disk, and image format; step output is a Python list of point cloud coordinates (x, y, z). The step GUI allows the user to interact with the image stack through a 3D scene and a 2D scene. The main segmentation tools are the point tool which adds single segmentation points; and the curve tool which uses splines to add a connected sequence of segmentation points. Plane orientation and position tools allow the user to translate and rotate the image plane so that segmentation can be carried out in any position, a feature not found in many segmentation software.

Hip Joint Centre Prediction Step. This step implements three methods for predicting the HJC from anatomical landmarks on the pelvis [11,1,10]. The step takes landmark name - landmark coordinates key-value pairs as input, and outputs the same with additional hip joint centre terms. The step GUI displays the landmarks in 3-D, allows the user to select the prediction method, patient population (male, female, or adult), and the landmarks to use for prediction.

The step was validated on 160 pelvis models segmented from CT images. Landmarks taken from the models were used to predict the HJC. The predicted HJC position was compared against a HJC position calculated as the centre of a sphere fitted to the model acetabulum. Mean distance between predicted and calculated HJCs was 12.9 ± 5.5 mm for the Tylkowski method, 7.3 ± 4.0 mm for the Bell method, and 5.7 ± 3.3 mm for the Seidel method.

Landmark Registration Step. This step registers a 3-D atlas femur mesh to femoral landmarks. The step uses rigid-body registration followed by non-rigid registration based on a femur shape model[14] to minimise the least-squares distance between known landmarks on the mesh and those given as input. A set of landmarks, a principal components model object (output by the PC Source step), and a femur parametric mesh (output by the Model Source step) are required as inputs; the registered mesh, the final transformation, and the registration error are given as outputs. The GUI displays the landmarks and femur mesh in 3-D, allows the selection of input landmarks that correspond to standard femoral landmarks, and displays the registration errors.

This step was validated on 12 femur meshes unseen to the shape model. Three landmarks were taken from each model (femoral head centre, medial epicondyle, and lateral epicondyle) and used to register the shape model's atlas model. Mean RMS distance between correspondent points on the actual and registered models was 4.3 ± 2.2 mm, which was roughly half of the error produced by rigid-body and isotropic scaling registration (8.5 ± 3.6 mm).

Point Cloud Registration Step. This step provides rigid-body and affine registration between two non-correspondent (using the iterative-closest point method [2]) or correspondent point clouds. The step takes as input two point clouds (source and target) and outputs the registered source point cloud along with the final transformation and the registration error. The GUI allows selection of the registration method, initial transformations, optimisation options, and displays the point clouds in 3-D.

Principal Components Fitting Step. This step fits an atlas surface mesh to a point cloud (and any provided landmarks) using rigid-body translation, rotation and deformation along its principal components[14]. The principal components ensures that the fitting is robust against noisy or incomplete point clouds. Step inputs are a point cloud, a mesh to be fitted, principal components, and optional initial transforms, data weights, and landmarks. Step outputs are the fitted mesh, final transformation, and fitting errors. The GUI displays the mesh, point cloud, and landmarks in 3-D, and allows the user to modify options regarding the fitting objective function and regularisation terms.

This step was validated on 30 femur meshes [14]. In leave-one-out experiments, the RMS surface-to-surface distance was calculated between each left out and fitted mesh. An average RMS error of 1.2 ± 0.3 mm was achieved using ten principal components, which was superior to the commonly-used host-mesh fitting method [6,5] (2.5 to 4.0 mm RMS).

Parametric Mesh Fitting Step. This step provides fine-scale fitting of parametric meshes to point clouds with Sobolev regularisation[8]. The step inputs are a parametric mesh to be fitted, a point cloud, and optional data weights. Step outputs are the fitted mesh and fitting errors. As above, the GUI allows the user to view the mesh and point cloud in 3-D, and to modify fitting options.

This step was validated by fitting a femur mesh to 30 segmented data clouds [14]. The entire fitting process involved the Point Cloud Registration step, the Principal Components Fitting step, and the Parametric Mesh Fitting step in series to progressively bring the mesh close to the data cloud. Thus, the following results represent a validation of not just this step, but of the latter half of the workflow presented here. For each femur, the RMS distance between data points and their closest point on the fitted mesh was calculated. Depending on the region of femur, the mean RMS distance was between 0.3 ± 0.07 mm and 0.61 ± 0.21 mm. This was below the resolution of the segmented images and that of most clinical images encountered in musculoskeletal modelling.

4   Summary

The MAP Client is an open-source workflow framework targeted toward musculoskeletal modelling. workflows can be quickly assembled in a graphical user interface from community created steps (plug-ins). Once created, workflows can be shared through the MAP Database along with data and run with minimal expert knowledge, enabling more researchers to participate in musculoskeletal modelling. As demonstrated in this paper, currently available steps enable the creation of patient-specific lower-limb bone models from imaging and mo-cap data. Development is under way to improve core MAP Client functionalities such as pipeline flow control, the user-interface, and integration with the MAP

Database. Also in development are steps for automatic segmentation and integration with popular biomechanics software. Our long-term goal is to foster a community of MAP users and developers to accelerate the clinical use of computational models in musculoskeletal modelling.

Acknowledgments. The authors are grateful to the Director and staff at the Victorian Institute of Forensic Medicine (VIFM) for providing the femur CT images. This work is funded by the US Food & Drug Administration (grant HHSF22320 1310119C).

References

1. Bell, A., Brand, R., Pedersen, D.: Prediction of hip joint centre location from external landmarks. Human Movement Science 8, 3–16 (1989)
2. Besl, P.J., McKay, N.D., Bed, P.J.: A method for registration of 3-d shapes. IEEE Transactions on Pattern Analysis and Machine Intelligence 14(2), 239–256 (1992)
3. Cleather, D.J., Bull, A.M.: The development of lower limb musculoskeletal models with clinical relevance is dependent upon the fidelity of the mathematical description of the lower limb. Part 2: patient-specific geometry. Proceedings of the Institution of Mechanical Engineers, Part H: Journal of Engineering in Medicine 226, 133–145 (2012)
4. Delp, S.L., Anderson, F.C., Arnold, A.S., Loan, P., Habib, A., John, C.T., Guendelman, E., Thelen, D.G.: OpenSim: Open-source software to create and analyze dynamic simulations of movement. IEEE Transactions on Biomedical Engineering 54, 1940–1950 (2007)
5. Fernandez, J.W., Hunter, P.J.: An anatomically based patient-specific finite element model of patella articulation: towards a diagnostic tool. Biomechanics and Modeling in Mechanobiology 4(1), 20–38 (2005),
   http://www.ncbi.nlm.nih.gov/pubmed/15959816
6. Fernandez, J.W., Mithraratne, P., Thrupp, S.F., Tawhai, M.H., Hunter, P.J.: Anatomically based geometric modelling of the musculo-skeletal system and other organs. Biomechanics and Modeling in Mechanobiology 2(3), 139–155 (2004),
   http://www.ncbi.nlm.nih.gov/pubmed/14685821
7. Fregly, B.J., Besier, T.F., Lloyd, D.G., Delp, S.L., Banks, S.A., Pandy, M.G., D'Lima, D.D.: Grand challenge competition to predict in vivo knee loads. Journal of Orthopaedic Research 30, 503–513 (2012)
8. Hunter, P.J., Nielsen, P.M., Smaill, B.H., LeGrice, I.J., Hunter, I.W., et al.: An anatomical heart model with applications to myocardial activation and ventricular mechanics. Critical Reviews in Biomedical Engineering 20(5-6), 403 (1992)
9. Larrabide, I., Omedas, P., Martelli, Y., Planes, X., Nieber, M., Moya, J.A., Butakoff, C., Sebastián, R., Camara, O., De Craene, M., Bijnens, B.H., Frangi, A.F.: GIMIAS: An open source framework for efficient development of research tools and clinical prototypes. In: Ayache, N., Delingette, H., Sermesant, M. (eds.) FIMH 2009. LNCS, vol. 5528, pp. 417–426. Springer, Heidelberg (2009)
10. Seidel, G.K., Marchinda, D.M., Dijkers, M., Soutas-Little, R.W.: Hip joint center location from palpable bony landmarks - A cadaver study. Journal of Biomechanics 28(8), 995–998 (1995)

11. Tylkowski, C.M., Simon, S.R., Mansour, J.M.: The Frank Stinchfield Award Paper. Internal rotation gait in spastic cerebral palsy. The Hip 89–125 (January 1982)

12. Yoo, T.S., Ackerman, M.J., Lorensen, W.E., Schroeder, W., Chalana, V., Aylward, S., Metaxas, D., Whitaker, R.: Engineering and algorithm design for an image processing api: a technical report on itk-the insight toolkit. Studies in Health Technology and Informatics, 586–592 (2002)

13. Yu, T., Lloyd, C.M., Nickerson, D.P., Cooling, M.T., Miller, A.K., Garny, A., Terkildsen, J.R., Lawson, J., Britten, R.D., Hunter, P.J., Nielsen, P.M.F.: The physiome model repository 2. Bioinformatics 27, 743–744 (2011)

14. Zhang, J., Malcolm, D., Hislop-Jambrich, J., Thomas, C.D.L., Nielsen, P.M.: An anatomical region-based statistical shape model of the human femur. Computer Methods in Biomechanics and Biomedical Engineering: Imaging & Visualization 1–10 (February 2014)

PGD-Based Model Reduction
for Surgery Simulation: Solid Dynamics
and Contact Detection

Carlos Quesada[1], Icíar Alfaro[1,2], David González[1,2], Elías Cueto[1,2],
and Francisco Chinesta[3]

[1] Aragón Institute of Engineering Research (I3A),
Universidad de Zaragoza, Zaragoza, Spain
[2] CIBER-BBN—Centro de Investigación Biomédica en Red en Bioingeniería,
Biomateriales y Nanomedicina, Zaragoza, Spain
[3] Ecole Centrale de Nantes, Nantes, France

Abstract. We present here an analysis of the possible advantages of
using *a priori* model order reduction techniques for real-time simulation
in computational surgery. Special attention will be paid to methods based
upon Proper Generalized Decomposition techniques. These techniques
allow for impressive savings in on-line computations by obtaining off-
line a reduced-order approach to the problem at hand. This approach
can be seen as a particular instance of meta-model, response surface or
—as we have coined it— a sort of *computational vademecum*. In this
work we detail the approach followed for the implementation of essential
aspects in computational surgery, such as solid dynamics and contact
detection.

1 Introduction

In order to fulfill real-time constraints, simulation techniques in the field of computational surgery has traditionally made use of the maximum amount of off-line
work possible. Classical pioneering works, in which soft living tissues were assumed linear elastic, usually pre-computed and stored stiffness matrices to avoid
assembling them on-line [7] [8]. However, it is clear that when considering non-
linear constitutive equations, and large-strain procedures (i.e., almost always),
it becomes increasingly difficult to pre-compute and store useful magnitudes so
as to speedup the on-line computation.

This work arises precisely from this question: *is it possible to pre-compute
everything?*. Of course, we must make clear what do we actually understand by
everything, but in any case, the answer will probably be "no". However, what if
we consider surgery as a parametric problem, in which every relevant parameter
could take a value within a prescribed interval? For instance, what if the position
of contact between scalpel and organ is a parameter of the model? What happens
if the magnitude and orientation of the force exerted by the surgeon during
surgery are considered as parameters? And what happens if the solution to this

F. Bello and S. Cotin (Eds.): ISBMS 2014, LNCS 8789, pp. 193–202, 2014.
© Springer International Publishing Switzerland 2014

parametric problem is obtained once for life, and only post-processed (instead of simulated) in real time? This would give rise to a sort of response surface or *computational vademecum* [6] that provides the solution to a given problem as a particularization of a high-dimensional, parametric model.

Of course, if we add more and more parameters to the problem, its complexity increases. And this increase is exponential with the number of parameters giving rise to the so-called *curse of dimensionality*. In words of Nobel prize winner R. B. Laughlin, "No computer existing, or that will ever exist, can break this barrier because it is a catastrophe of dimension" [14].

However, nowadays we know how to approximately solve this problem. Of course, it is mandatory to perform some kind of model reduction, i.e., to keep the number of degrees of freedom to a minimum by a judicious choice of the subspace spanning the solution of the problem. But also to construct this optimal subspace, which is principle high-dimensional, in an efficient way. These two ingredients are fulfilled by a technique coined as Proper Generalized Decomposition (PGD) [13] [2] [5].

We describe here how to exploit the characteristics of the PGD method to solve high-dimensional parametric problem arising in the real-time (haptic) simulation of surgery procedures. To this end, in Section 2 we recall the basics of PGD.

2   A Brief Overview of Proper Generalized Decomposition

The main ingredient of the method here reported is the intensive use of computational vademecums [6] generated off-line. To begin with, consider the case of a vademecum in which we would like to store the displacement field $u(x)$ of a non-linear (here, hyperelastic) organ Ω under the action of a surgical tool force (assumed for the sake of simplicity as unitary and always acting in the vertical direction) at any point s of its boundary region $\bar{\Gamma} \subset \Gamma_t \subset \Gamma = \partial\Omega$. This renders a problem defined in general in $\mathbb{R}^5$ ($u = u(x, s)$), although if s is interpolated using nearest-neighbour interpolation, it can be seen as a one-dimensional parameter (the node in which the load is acting), rendering a problem in $\mathbb{R}^4$.

As usual in the FE method, we consider now the weak form of the equilibrium equations (balance of linear momentum) without inertia terms. The interested reader can consult [10] for a detailed explanation on the dynamic case. The (doubly-) weak form of the problem, extended to the whole geometry of the organ, Ω and the portion of its boundary which is accessible to load during surgery, $\bar{\Gamma} \subset \Gamma_t$, consists in finding the displacement $u \in \mathcal{H}^1$ such that for all $u^* \in \mathcal{H}_0^1$:

$$\int_{\bar{\Gamma}} \int_{\Omega} \boldsymbol{\nabla}_s u^* : \boldsymbol{\sigma} d\Omega d\bar{\Gamma} = \int_{\bar{\Gamma}} \int_{\Gamma_{t2}} u^* \cdot t d\Gamma d\bar{\Gamma} \tag{1}$$

where $\Gamma = \Gamma_u \cup \Gamma_t$ represents the boundary of the organ, divided into essential and natural regions, and where $\Gamma_t = \Gamma_{t1} \cup \Gamma_{t2}$, i.e., regions of homogeneous and non-homogeneous, respectively, natural boundary conditions. Here, $t = -e_k \cdot \delta(x - s)$, where δ represents the Dirac-delta function and e_k the unit vector along the z-coordinate axis (we consider here, as mentioned before, and for the ease of exposition, a unit load directed towards the negative z axis of reference).

The key aspect of the method here proposed is that PGD techniques efficiently construct the computational vademecum $u(x, s)$ by constructing, in an iterative way, an approximation to the solution in the form of a finite sum of separable functions [5]. If we assume that the method has converged to a solution, at iteration n of this procedure,

$$u_j^n(x, s) = \sum_{k=1}^{n} X_j^k(x) \cdot Y_j^k(s), \tag{2}$$

where the term u_j refers to the j-th component of the displacement vector, $j = 1, 2, 3$ and functions $X^k(x)$ and $Y^k(s)$ represent the separated functions used to approximate the unknown field, obtained in previous iterations of the PGD algorithm. At this stage, the objetive of PGD is to provide the solution with an improvement given by the $(n + 1)$-th term of the approximation,

$$u_j^{n+1}(x, s) = u_j^n(x, s) + R_j(x) \cdot S_j(s), \tag{3}$$

where $R(x)$ and $S(s)$ are the sought functions that improve the approximation. In an equivalent manner, admissible variations of this displacement field will be given by

$$u_j^*(x, s) = R_j^*(x) \cdot S_j(s) + R_j(x) \cdot S_j^*(s).$$

Of course, the price to pay during this procedure is that, even if the original problem is linear, PGD needs for the solution of a non-linear problem, i.e., to determine a product of functions, see Eq. (3). The reader can think of any linearization method available in the literature. However, practical examples of implementation can be found at [15].

3   Quasi-Static Palpation of the Liver

As an example, we consider here the case of liver palpation. Liver geometry has been obtained from the SOFA project [1] and post-processed in order to obtain a mesh composed by 8559 nodes and 10519 tetrahedra.

We have assumed a Saint Venant-Kirchhoff hyperelastic model, with Young's modulus of 160 kPa, and a Poisson coefficient of 0.48, thus nearly incompressible [8]. The $\bar{\Gamma}$ surface, where the load can be located, has been defined as the whole boundary of the domain, even if in this case, only the frontal part of the organ is usually accessible to the surgeon. This region includes 2009 of the 8559 nodes of the model.

Model's solution was composed by a total of $n = 167$ functional pairs $X_j^k(x) \cdot Y_j^k(s)$ (see Eq. (2)). The third component (i.e., $j = 3$) of the first three modes $X_3^k(x)$ is depicted in Fig. 1. The same is done in Fig. 2 for functions Y, although in this case they are defined only on the boundary of the domain, i.e., $\bar{\Gamma} = \partial\Omega$.

Feedback rates in the order of 1 kHz are obtained without problems for a Touch haptic device [9], see Fig. 3.

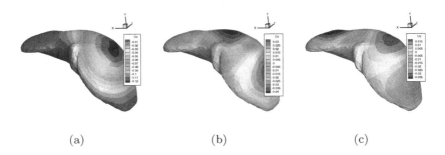

Fig. 1. Three first functions $X_3^k(\boldsymbol{x})$, $k = 1, 2, 3$, for the simulation of the liver

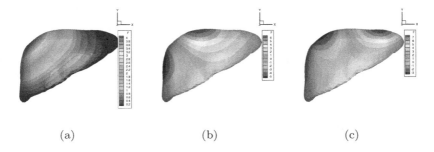

Fig. 2. Three first functions $Y_3^k(\boldsymbol{s})$, $k = 1, 2, 3$, for the simulation of the liver. Note that, in this case, functions $\boldsymbol{Y}^k(\boldsymbol{s})$ are defined on the boundary of the liver only.

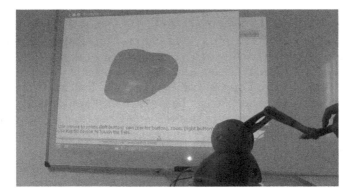

Fig. 3. Implementation of liver palpation with haptic feedback on a HP ProBook 6470b laptop equipped with a Geomagic Touch device

4 Interactive Non-linear Solid Dynamics

Pre-computing and storing in memory data relative to interactive solid dynamics of non-linear solids is far from being trivial. In [10] we presented a method based on the construction of a computational vademecum whose parameters were precisely initial values (displacement and velocity) of the problem. In this way, a sort of black-box integrator was made such that, once nodal displacement and velocity values were known at time t, they could be considered as initial values for the subsequent time step $t + \triangle t$.

Consider the weak form of the solid dynamics equations, i.e.: given $\boldsymbol{f}, \boldsymbol{g}, \boldsymbol{h}, \boldsymbol{u}_0$ and $\dot{\boldsymbol{u}}_0$ find $\boldsymbol{u}(t) \in \mathcal{S}_t = \{\boldsymbol{u}|\boldsymbol{u}(\boldsymbol{x},t) = \boldsymbol{g}(\boldsymbol{x},t), \ \boldsymbol{x} \in \Gamma_u, \ \boldsymbol{u} \in \mathcal{H}^1(\Omega)\}, \ t \in [0,T]$, such that for all $\boldsymbol{w} \in \mathcal{V}\{\boldsymbol{u}|\boldsymbol{u}(\boldsymbol{x},t) = \boldsymbol{0}, \ \boldsymbol{x} \in \Gamma_u, \ \boldsymbol{u} \in \mathcal{H}^1(\Omega)\}$,

$$(\boldsymbol{w}, \rho\ddot{\boldsymbol{u}}) + a(\boldsymbol{w}, \boldsymbol{u}) = (\boldsymbol{w}, \boldsymbol{f}) + (\boldsymbol{w}, \boldsymbol{h})_{\Gamma_t} \tag{4a}$$

$$(\boldsymbol{w}, \rho\boldsymbol{u}(0)) = (\boldsymbol{w}, \rho\boldsymbol{u}_0) \tag{4b}$$

$$(\boldsymbol{w}, \rho\dot{\boldsymbol{u}}(0)) = (\boldsymbol{w}, \rho\dot{\boldsymbol{u}}_0). \tag{4c}$$

The main ingredient of the developed computational vademecum is to express the displacement field as a parametric field such as $\boldsymbol{u} : \bar{\Omega} \times]0,T] \times \mathcal{I} \times \mathcal{J} \to \mathbb{R}^3$, where $\mathcal{I} = [\boldsymbol{u}_0^-, \boldsymbol{u}_0^+]$ and $\mathcal{J} = [\dot{\boldsymbol{u}}_0^-, \dot{\boldsymbol{u}}_0^+]$ represent the considered intervals of variation of initial boundary conditions, $\boldsymbol{u}_0$ and $\dot{\boldsymbol{u}}_0$. This makes it necessary to define a new (triply-) weak form where:

$$a(\boldsymbol{w}, \boldsymbol{u}) = \int_{\mathcal{I}} \int_{\mathcal{J}} \int_{\Omega} \boldsymbol{\nabla}^s \boldsymbol{w} : \mathbf{C} : \boldsymbol{\nabla}^s \boldsymbol{u} \ d\Omega d\dot{\boldsymbol{u}}_0 d\boldsymbol{u}_0,$$

$$(\boldsymbol{w}, \boldsymbol{f}) = \int_{\mathcal{I}} \int_{\mathcal{J}} \int_{\Omega} \boldsymbol{w}\boldsymbol{f} \ d\Omega d\dot{\boldsymbol{u}}_0 d\boldsymbol{u}_0,$$

$$(\boldsymbol{w}, \boldsymbol{h})_{\Gamma} = \int_{\mathcal{I}} \int_{\mathcal{J}} \int_{\Gamma_t} \boldsymbol{w}\boldsymbol{h} \ d\Gamma d\dot{\boldsymbol{u}}_0 d\boldsymbol{u}_0.$$

In order to solve such a high-dimensional problem (a straightforward approach by meshing the entire parametric space would not be possible due to the so-called curse of dimensionality, arising to an enormous number of degrees of freedom), PGD methods construct an approximation to the solution as a finite sum of separable functions,

$$\boldsymbol{v}^h(\boldsymbol{x}, t, \boldsymbol{u}_0, \dot{\boldsymbol{u}}_0) = \left[\sum_{i=1}^{N} \boldsymbol{F}_i(\boldsymbol{x}) \circ \boldsymbol{G}_i(\boldsymbol{u}_0) \circ \boldsymbol{H}_i(\dot{\boldsymbol{u}}_0)\right] \circ \boldsymbol{d}(t), \tag{5}$$

where the nodal coefficients $\boldsymbol{d}(t)$ carry out all the time-dependency of the solution and the symbol "$\circ$" stands for the entry-wise Hadamard or Schur multiplication of vectors. PGD computes these separable functions by first linearizing the non-linear problem (since we seek for a product of functions) by employing your favorite linearization technique (usually fixed-point algorithms, but also Newton or quasi-Newton methods are equally possible). PGD computes one sum at a time, then one product at a time, see [6], for instance.

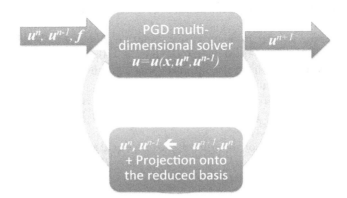

Fig. 4. Sketch of the proposed algorithm. A POD basis is employed for the parametrization of the space of initial conditions of the problem (for forward Euler schemes, for instance, this means in practice that the solution deepens on the displacement at time steps n and $n-1$, respectively).

The very last detail in the implementation is to search for an approximation not for the whole time interval of the problem, $]0, T]$, but for $]0, \triangle t]$:

$$\boldsymbol{v} : \bar{\Omega} \times]0, \triangle t] \times \mathcal{I} \times \mathcal{J} \times [h^-, h^+] \to \mathbb{R}^3,$$

where $\triangle t$ represents the necessary time to response prescribed by the particular envisaged application. For instance, for haptic feedback it has been already mentioned that a physical sensation of touch needs for some 500 Hz to 1 kHz feedback rate. This means that $\triangle t = 0.001$ seconds. This value $\triangle t$ is not the necessary time step to achieve stability in the time integration chosen (that can be smaller if needed), although it can be coincident (and will be for all the examples shown hereafter).

But the space of initial conditions, if approximated by finite elements, needs for a very large number of degrees of freedom (three velocity components and three displacements per node of the model). Our approach includes the projection of this subspace into another one, conveniently reduced by employing POD techniques. The resulting technique, see Fig. 4 is a sort of black-box integrator that, given the conditions at time step t as initial conditions, provides the method with the resulting displacement and velocity fields at time $t + \triangle t$.

Simulating a palpation, a ramp load of 5 N is applied at a particular point of the liver surface during a period of 0.25 seconds, and then released during other 0.25 seconds. The liver is then left vibrating free. Even if the liver tissue is well-known to posses some kind of viscoelastic properties, these have been neglected. The purpose of this example is not to obtain an extremely realistic simulation from a physiological point of view, but to show the performance of the technique. In particular, the influence of the number of modes chosen to parametrize the space of initial conditions on the long-term behaviour of the solution. To this end, a reference solution has been computed by employing an

HHT time integrator [11] and standard finite elements. POD modes have been extracted from this reference solution to construct the basis for the combined PGD-POD integrator.

It can be noticed that, in the idealized situation of absence of any type of damping, after the load release, the liver continues vibrating indefinitely. A PGD-POD solution has been computed by employing 1, 3 and 7 modes in the basis of the space of initial conditions. It can be noticed, from Fig. 5 that increasing the number of modes, as expected, provides converging results towards the reference solution. In fact, it can be noticed how the reduction in the number of degrees of freedom, very much like in classical model order reduction, eliminates high frequencies from the solution (those with the lowest content in energy) and therefore oscillations around the reference solution can be observed. The richer the basis is, the smaller the amplitude of these oscillations and the smaller their period.

5   Contact Simulation under the Vademecum Approach

One of the most popular families of real-time contact detection algorithms is based on the use of distance fields [4] [3]. Distance fields (level sets) constitute a very convenient way of representing very intricate geometries for contact detection, but has been traditionally considered as non-apt for real-time contact simulation between deformable solids, since the distance field must be updated according to the deformation of one of the solids [16].

The method here developed assumes that a computational vademecum $u(x, s)$ has already been computed for the solids under consideration. As mentioned before, this multi-dimensional solution provides a general solution for the displacement field in the solids under an arbitrary load acting on $\bar{\Gamma}$.

The collision detection method assumes that one of the solids, say Ω_2, is modeled as a *pointshell* [3], i.e., a set of boundary points (assumed for simplicity identical to the boundary nodes of the solid model, although this is not strictly necessary) equipped with normals to the surface. As such, the algorithm is not symmetric, since the choice of which body is chosen to be equipped with the pointshell affects the final result of the simulation, albeit slightly. The other solid, Ω_1, is in turn equipped with a signed distance field, see Fig. 6.

In our implementation, following closely [3], at every haptic cycle, contact penalty forces are determined by querying the points of Ω_2 against the distance field associated to Ω_1. Traditionally, this approach has been considered not valid for haptic feedback requirements, if we deal with two deformable solids, see [16]. This limitation is due to the need of updating the distance field along with the deformation of Ω_1, which is not an easy task, even if the collision detection loop is not performed at every haptic cycle, as in [12]. In the approach here presented, there is no need for such an update, since a high-dimensional distance field is computed off-line that contains the distance fields for any deformed configuration of the solid.

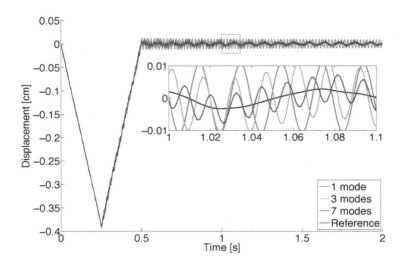

Fig. 5. Response of the liver for a peak load. Reference (FE) results, and PGD results with basis composed by 1, 3 and 7 modes.

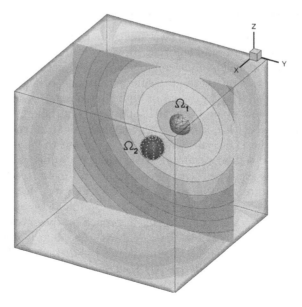

Fig. 6. Sketch of the collision detection algorithm. Solid Ω_1 is equipped with a distance field, represented in the Figure, while solid Ω_2 is represented, for collision detection purposes, as a pointshell, i.e., a collection of boundary nodes and normals to the surface (not represented for simplicity).

Once collision has been detected, a force $F = -k_c dn$ is applied to both solids, provided that d is trilinearly interpolated from the distance field accompanying Ω_1. k_C is the contact penalty stiffness. The minus sign assumes that normals to Ω_2 point towards the interior of the solid, although this is completely arbitrary. Finally, n represents the normal to Ω_2 in the deformed configuration. An equal force is applied to Ω_1, regardless of its geometry, since we have no information on its normals to the surface. This very simple algorithm preserves continuity in the force computation, essential to perceive haptic response as realistic.

6   Conclusions

In this paper we analyze how the traditional idea of pre-computing off-line as much as possible can be updated. Based on the concept of computational vademecum, our implementation of a haptic surgical simulator employs an off-line solution to a high-dimensional parametric problem. As a preliminar example, only the contact position of the surgical tool and the organ is considered as a parameter, for the sake of simplicity in the exposition. This concept is then extended to dynamics and contact detection under the same rationale, satisfactorily. What is remarkable in this approach is the possibility of employing implicit, energy and momentum conserving schemes for the simulation of solid dynamics. This ensures very high rates of accuracy and energy conservation even for long times of simulation, without he well-known numerical dissipation of classical explicit integration schemes. We strongly believe that the extension of the techniques here presented opens interesting possibilities in the field.

References

1. Allard, J., Cotin, S., Faure, F., Bensoussan, P.-J., Poyer, F., Duriez, C., Delingette, H., Grisoni, L.: SOFA an Open Source Framework for Medical Simulation. In: Medicine Meets Virtual Reality (MMVR'15), Long Beach, USA (February 2007)
2. Ammar, A., Mokdad, B., Chinesta, F., Keunings, R.: A new family of solvers for some classes of multidimensional partial differential equations encountered in kinetic theory modeling of complex fluids. J. Non-Newtonian Fluid Mech. 139, 153–176 (2006)
3. Barbič, J., James, D.L.: Six-dof haptic rendering of contact between geometrically complex reduced deformable models. IEEE Transactions on Haptics 1(1), 39–52 (2008)
4. Bridson, R., Marino, S., Fedkiw, R.: Simulation of clothing with folds and wrinkles. In: Proceedings of the 2003 ACM SIGGRAPH/Eurographics Symposium on Computer Animation, SCA 2003, pp. 28–36. Eurographics Association, Aire-la-Ville (2003)
5. Chinesta, F., Ammar, A., Cueto, E.: Recent advances in the use of the Proper Generalized Decomposition for solving multidimensional models. Archives of Computational Methods in Engineering 17(4), 327–350 (2010)
6. Chinesta, F., Leygue, A., Bordeu, F., Aguado, J.V., Cueto, E., Gonzalez, D., Alfaro, I., Ammar, A., Huerta, A.: PGD-Based Computational Vademecum for Efficient Design, Optimization and Control. Archives of Computational Methods in Engineering 20(1), 31–59 (2013)

7. Cotin, S., Delingette, H., Ayache, N.: Real-time elastic deformations of soft tissues for surgery simulation. In: Hagen, H. (ed.) IEEE Transactions on Visualization and Computer Graphics, vol. 5(1), pp. 62–73. IEEE Computer Society (1999)

8. Delingette, H., Ayache, N.: Soft tissue modeling for surgery simulation. In: Ayache, N., Ciarlet, P. (eds.) Computational Models for the Human Body. Handbook of Numerical Analysis, pp. 453–550. Elsevier (2004)

9. Geomagic. OpenHaptics Toolkit. 3D systems - Geomagic solutions, 430 Davis Drive, Suite 300 Morrisville, NC 27560 USA (2013)

10. Gonzalez, D., Cueto, E., Chinesta, F.: Real-time direct integration of reduced solid dynamics equations. Internatinal Journal for Numerical Methods in Engineering (accepted, 2014)

11. Hilber, H.M., Hughes, T.J.R., Taylor, R.L.: Improved numerical dissipation for time integration algorithms in structural dynamics. Earthquake Engineering and Structural Dynamics 5, 283–292 (1977)

12. Jeřábková, L., Kuhlen, T.: Stable cutting of deformable objects in virtual environments using xfem. IEEE Comput. Graph. Appl. 29(2), 61–71 (2009)

13. Ladeveze, P.: Nonlinear Computational Structural Mechanics. Springer, N.Y (1999)

14. Laughlin, R.B., Pines, D.: The theory of everything. Proceedings of the National Academy of Sciences 97(1), 28–31 (2000)

15. Niroomandi, S., González, D., Alfaro, I., Bordeu, F., Leygue, A., Cueto, E., Chinesta, F.: Real-time simulation of biological soft tissues: a pgd approach. International Journal for Numerical Methods in Biomedical Engineering 29(5), 586–600 (2013)

16. Teschner, M., Kimmerle, S., Heidelberger, B., Zachmann, G., Raghupathi, L., Fuhrmann, A., Cani, M.-P., Faure, F., Magnenat-Thalmann, N., Strasser, W., Volino, P.: Collision detection for deformable objects. Computer Graphics Forum 24(1), 61–81 (2005)

Bender: An Open Source Software for Efficient Model Posing and Morphing

Julien Finet[1], Ricardo Ortiz[1], Johan Andruejol[1], Andinet Enquobahrie[1], Julien Jomier[1], Jason Payne[2], and Stephen Aylward[1]

[1] Kitware, Inc.
[2] Air Force Research Laboratory

Abstract. In this paper, we present Bender, an interactive and freely available software application for changing the pose of anatomical models that are represented as labeled, voxel-based volumes.

Voxelized anatomical models are used in numerous applications including the computation of specific absorption rates associated with cell phone transmission energies, radiation therapy, and electromagnetic dosimetry simulation. Other applications range from the study of ergonomics to the design of clothing. Typically, the anatomical pose of a voxelized model is limited by the imaging device used to acquire the source anatomical data; however, absorption of emitted energies and the fit of clothes will change based on anatomic pose.

Bender provides an intuitive, workflow-based user-interface to an extensible framework for changing the pose of anatomic models. Bender is implemented as a customized version of 3D Slicer, an image analysis and visualization framework that is widely used in the medical computing research community. The currently available repositioning methods in Bender are based on computer-graphics techniques for rigging, skinning, and resampling voxelized anatomical models. In this paper we present the software and compare two resampling methods: a novel extension to dual quaternions and finite element modeling (FEM) techniques. We show that FEM can be used to quickly and effectively resample repositioned anatomic models.

1 Introduction

The driving application for the work in this paper is the use of anatomical models in numerical simulations to characterize the absorption of radio frequency (RF) energy within tissues, and the associated temperature responses. The specific absorption rates throughout a body exposed to directed energies will change based on the body's anatomic pose. Acquiring anatomical body models in various postures and for various body types can be problematic due to medical scanner costs, post-processing labor efforts and acquisition constraints on pose. Typically, subjects must be lying down during x-ray computed tomography (CT) and magnetic resonance imaging (MRI) acquisitions, and therefore anatomical image datasets are mostly constrained to this pose. Repositioning systems that

F. Bello and S. Cotin (Eds.): ISBMS 2014, LNCS 8789, pp. 203–210, 2014.

are tailored to a specific voxelized model do exist and provide accurate repositions of that single model [3,4]. These costly systems usually do not provide any method for repositioning other voxelized models. Researchers have also defined algorithms whereby precise anatomical mechanics can be derived from a voxelized model and used to reposition that model. These systems, however, require the specification of muscle and tendon connections and precise bone-joint modelling [5]. The level of detail required is beyond what is readily available and a significant operator time is required to define joint and muscle kinematics for these models as input to these algorithms [1,2]. Vaillant et al [14] developed geometric method that handles skin contact effects and muscular bulges in real-time. Kavan et al [15] have developed a skinning technique based on the concept of joint-based deformers. Mohr et al [16] presented a technique that uses examples that fit the parameters of a deformation model that best approximates the original data.

There is a need for open-source, freely available, and extensible software that can generate a new voxelized model that represents the original model resampled into the new position accurately and with only modest requirements regarding user effort and expertise. This paper introduces and evaluates such software.

2   Methods

Bender is an open-source toolkit based on 3D Slicer [8] that provides algorithms and a user-friendly application for repositioning voxelized anatomical models into a desired pose [9]. It is a novel extension to computer graphics methods for rigging, skinning and posing to work with voxelized volumes [11]. The three main technical contributions presented in this paper are (1) the development of an easy-to-use workflow-based interface, (2) the incorporation of existing motion-capture database into the model-repositioning workflow, and (3) the development and evaluation of novel anatomical model resampling techniques.

2.1   Workflow

A workflow paradigm was chosen for Bender to provide an intuitive interface that guides users through the complexities of anatomic model repositioning. The steps of that workflow are as follows:

1. Rigging: This involves specifying a skeleton that represents the linear sections and joints of the body, by which the body will be repositioned (Fig. 1(a)). There are two options available to the user to define the rig: (1) they may click and drag in 2D or 3D views to place and edit each joint respectively or (2) they may modify by dragging the joints of an existing rig to fit the volume. Such task is relatively fast, especially with option (2); only a few minutes of work is required for this step.
2. Skinning: This is a 3D painting process in which the bones, soft tissues, and skin that should be moved with each rig section are explicitly associated

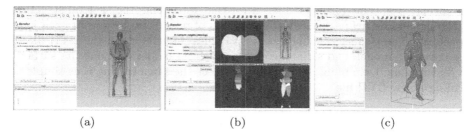

(a)   (b)   (c)

Fig. 1. (a) Rigging: Manually create or load an armature to fit on the volume bones. (b) Skinning: Associate a unique bone index for each anatomical voxel. (c) Posing: Reposition the anatomical model with a pose defined by the armature.

with each section (Fig. 1(b)). A default skinning map is generated by a heat diffusion algorithm on the rig using the underlying volume. Hinge joints, such as the elbow and knee, do not require much editing, however the ball and socket joints, i.e. the shoulder and hip joints, can be labor intensive. A few hours could be needed for this step. Note that the rigging and skinning are both one-time tasks for a given voxel volume.

3. Posing: The rigging is bent at its joints to define the target repositioning of the body.
4. Resampling: The bones, soft tissues, and skin in the voxelized model are resampled onto the repositioned rigging to create a new voxelized model (Fig. 1(c))

2.2   Incorporation of Existing Motion-Capture Data Libraries

The posing task is achieved by the user clicking and dragging the rig to define joint rotations. Experience can be improved by adding rotation constraints at the joints. However, the posing step would still remain highly user-dependent. Specifying a pose that is mechanically feasible and that looks natural is a tedious task that requires extensive user interaction.

To simplify the posing process, we developed a pipeline to load predefined poses that are widely available over the web. Typically these offer outstanding realism because they are created using human subjects and motion capture devices. The poses in these databases are typically specified using the Biovision hierarchy (BVH)

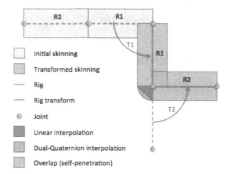

Fig. 2. Comparison between linear and dual quaternion interpolation techniques. While volume is more preserved in the elbow area with DQ interpolation, volume is lost in both cases in the forearm-arm junction due to lack of collision detection and force response.

file format [12]. This format defines a rig that represents the skeletal structure of the model and defines poses and sequences of poses as rotations at each joint in the rig. The Bender pipeline reads BVH files, fits the rig in the BVH file to the voxelized model, and skins the entire voxelized model (not only its surface) to that rigging. Definining a new pose that uses the same rigging is as simple as loading a new BVH file.

2.3   Advanced Anatomical Model Resampling Techniques

Bender incorporates advanced methods for resampling voxelized models, after rigging, skinning, and repositioning have been specified. The first resampling method is based on a linear resampling technique. For each voxel of the input models, the joints transforms (i.e. poses) are linearly combined using the weights derived from the skinning map. And each input voxel value is copied into the repositioned model at the voxel transformed coordinates. Linearly resampled models are typically not realistic as they do not take into account tissue deformation properties. The second resampling method is based on a Dual Quaternion technique for tissue deformation approximation [13]. Quaternion Spherical Linear Interpolation (Slerp) ensures constant-speed motion and improves the realism of the estimated deformation. This method is an improvement over the linear method; however, unrealistic tissue deformations can still result (see Fig. 2). To improve the resampling realism, particularly regarding self intersections, we implemented a Finite Element Method (FEM)-based resampling technique that takes into account tissue-specific deformation properties. FEM is a numerical technique for finding approximate solutions to boundary value problems for differential equations. Bender's FEM methods make use of the Simulation Open Framework Architecture (SOFA) [10] toolkit.

3   Experiments and Results

Experiments were conducted to evaluate the performance of Bender on arm repositioning. The experiments include quantitative and qualitative analysis to compare and contrast the different resampling techniques. The arm dataset was extracted from the Visible Man dataset and three subsampled volumes (2mm, 3mm and 4mm) were generated. The Bender workflow was then applied to each volume by following the same rigging and skinning steps. For the resampling step, we evaluated four techniques: linear interpolation(LN), dual quaternion interpolation (DQ), FEM without intersection (aka. collision) detection (FEMw/oC) and FEM with collision detection (FEMw/C). Four rotation angles were applied to the rig at the elbow joint ($0°$, $45°$, $60°$ and $90°$). For the FEM techniques, two extra steps were required: the voxelized model was tetrahedralized to generate a multi-material mesh and once that tetrahedral mesh was resampled, it was re-voxelized to obtain the final repositioned voxelized model. Three metrics were used to evaluate the techniques:

1. **Computational time** required to perform the repositioning.
2. **Volume preservation** of the total input volume was assessed. The metric can be obtained by calculating the ratio between the number of non-air voxels in the input dataset and the repositioned model. The ratio was also calculated for each individual tissue. A ratio of 0% signifies that there was no volume loss. A positive ratio means that volume was added through the repositioning process, and a negative ratio means that volume was lost.
3. A **qualitative study** was done to visually compare the results. We generated screenshots of the 3D rendered volumes for the general deformation and 2D reformatted slice views for the local changes and have subjects review and score the results.

Computational Time: For all techniques, computational time is naturally proportional to the image size (Fig. 3(a)). Downsampling the dataset significantly reduces running time (Fig. 3(b)). The LN and DQ techniques behave differently than the FEM techniques regarding the rotation angles. Running time for LN&DQ techniques is relatively stable at any angle. For large rotations ($> 60°$), processing time is increased due to the number of voxels that must be filled by the interpolation techniques. Self-penetration does not impact the overall computation time as penetration is not detected by the LN&DQ techniques. For the FEM techniques, three trends can be observed. First, as the angle increases, computation time increases due to large deformations in the soft tissues; more steps are required to rotate the model. Second, as size of the image decreases, computation time is reduced. The voxelization of the posed tetrahedral mesh is performed by browsing through each voxel of the final image and calculating the label value of the voxel in the original input voxelized model. Lastly, the surface-based collision detection for the FEMw/C technique is computationally expensive. As a result, the FEM techniques are 20 to 30 times slower than the interpolation techniques. Experiments were run on a 2.33GHz 8-core 64b CPU desktop machine. All tested techniques are currently single threaded and do not use the GPU. We believe there will be a significant improvement in running the experiments on a multi-threaded implementation.

Volume Preservation: The total volume and individual organ volume changes (e.g. fat volume change) follow the same trends (see Fig. 4). Generally, the FEM techniques preserve volume significantly better than the LN&DQ interpolation techniques. As expected, the Dual Quaternion technique has slightly better results than the linear technique. The FEM techniques with or without collisions produce similar results for small rotation angles, when there is no self-penetration. However the FEM with collision preserves volume better than the FEM without collision. Two limitations to this metric should be noted: firstly some living tissues can be heterogeneous and non perfectly incompressible and secondly the overall volume change may not accurately represent local changes. For example, for wide angles, volume is added by the DQ interpolation in the elbow region. However, volume is lost by the self-penetration. Further investigation of the FEM techniques will be needed in the future because volume is still being lost even when there is no self-penetration. This might be due to the FEM formulation.

Qualitative Metric: The interpolation techniques and the FEM techniques results show significant difference in deformation (see Figures 5 and 6). However, within each category, only small differences can be noted. For example, except in the elbow area, the rest of the arm shows little difference for the LN and DQ techniques. Similarly, significant difference is observed in the elbow region for the FEM techniques. For the interpolation techniques, the DQ interpolation produces more realistic repositioned models in the elbow area than the linear interpolation. For the FEM techniques, the FEM with collision detection produces more realistic deformations even for small angles than the FEM without collision. Nonetheless, this comes at the expense of having to compute the self-collision interactions.

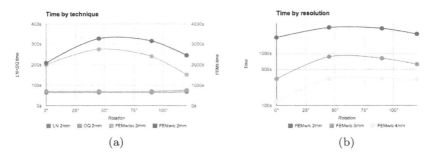

Fig. 3. (a) Time by technique (b) Time by resolution

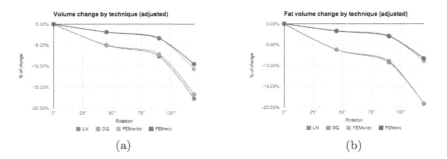

Fig. 4. (a) Volume change by technique for all tissues and (b) for fat tissue only

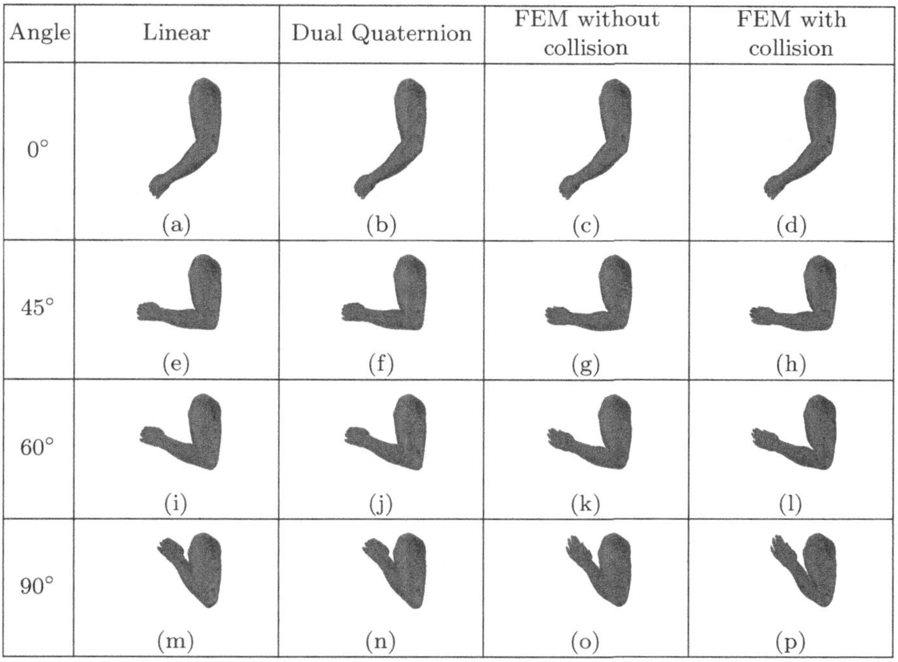

Angle	Linear	Dual Quaternion	FEM without collision	FEM with collision
0°	(a)	(b)	(c)	(d)
45°	(e)	(f)	(g)	(h)
60°	(i)	(j)	(k)	(l)
90°	(m)	(n)	(o)	(p)

Fig. 5. 3D rendered visualizations of different arm repositioning techniques. The main difference resides in the anticubital area of the elbow.

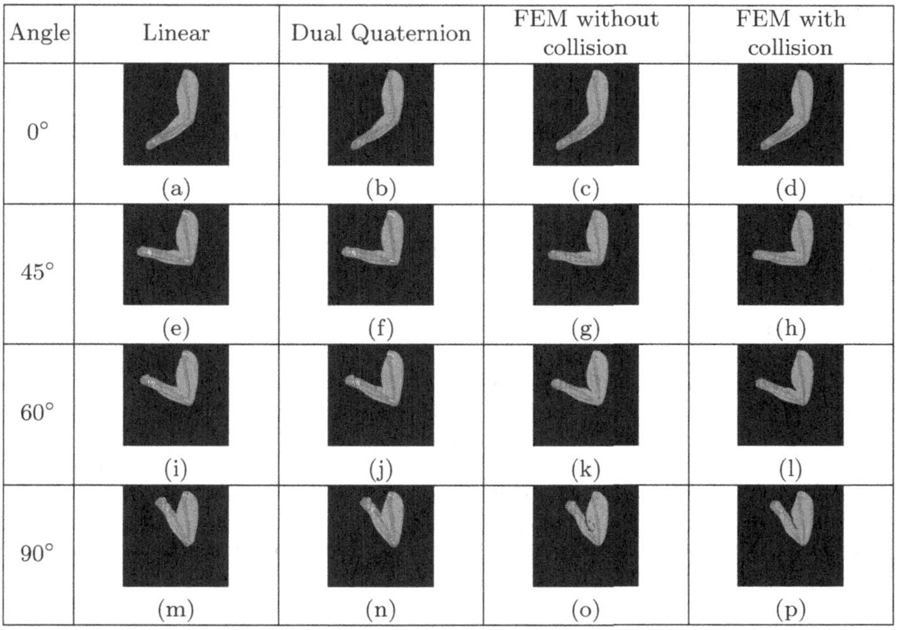

Angle	Linear	Dual Quaternion	FEM without collision	FEM with collision
0°	(a)	(b)	(c)	(d)
45°	(e)	(f)	(g)	(h)
60°	(i)	(j)	(k)	(l)
90°	(m)	(n)	(o)	(p)

Fig. 6. 2D views of different arm repositioning techniques

4 Conclusion

We introduced, an interactive, software application for repositioning voxelized anatomical models. Bender is released as an open-source software that is cross-platform and freely available. The software provides a user-friendly workflow-based module for rigging, skinning and resampling voxelized anatomical models. Quaternion and finite element based techniques were developed and evaluated to resample the bones, soft tissues, and skin of the voxelized model onto the repositioned rigging. Future work includes improving the finite element formulation, the overall computational time and evaluation of results.

Acknowledgments. The development and evaluation of Bender has been supported, in part, by the AFRL SBIR FA8650-13-M-6444 and by the NIH grant R44OD018334.

References

1. Aubel, A., Thalmann, D.: MuscleBuilder: A modeling tool for human anatomy. Journal of Computation Science and Technology 19(5), 585–595 (2004)
2. Wilhelms, J., Van Gelder, A.: Anatomically Based Modeling. In: Proc. of the 24th Annual Conf. on Comp. Graph. and Interactive Tech., pp. 173–180 (August 1997)
3. Remcom: VariPose (2012), http://www.remcom.com/varipose
4. SEMCAD-X: Poser, http://www.speag.com/products/semcad
5. Bhutani, R., Sharma, S.: Thesis: Repositioning of Human Body Models. Department of Mechanical Engineering, IIT Delhi, India (2010)
6. Rohmer, D., Hahmann, S., Cani, M.-P.: Exact volume preserving skinning with shape control. In: Eurographics/ACM SIGGRAPH Symp. on Comp. Animation (2009)
7. Segars, W.P., Sturgeon, G.: 4D XCAT phantom for multimodality imaging research. Medical Physics 37(9), 4902–4915 (2010)
8. Fedorov, A., Beichel, R., Kalpathy-Cramer, J., Finet, J., Fillion-Robin, J.-C., Pujol, S., Bauer, C., Jennings, D., Fennessy, F., Sonka, M., Buatti, J., Aylward, S.R., Miller, J.V., Pieper, S., Kikinis, R.: 3D Slicer as an Image Computing Platform for the Quantitative Imaging Network. Magn Reson Imaging 30(9), 1323–1341 (2012), PMID: 22770690
9. Bender, http://public.kitware.com/Wiki/Bender
10. SOFA: Simulation Open Framework Architecture, http://www.sofa-framework.org
11. Magnenat, N., Laperrière, R., Thalmann, D.: Joint-dependent Local Deformations for Hand Animation and Object Grasping. In: Proc. on Graph. Intf., pp. 26–33 (1988)
12. Meredith, M., Maddock, S.: Motion Capture File Formats Explained. Department of Computer Science, University of Sheffield (2001)
13. Kavan, L., Collins, S., Žára, J., O'Sullivan, C.: Skinning with Dual Quaternions. In: Proc. Symp. on Interactive 3D Graphics and Games, pp. 39–46 (2007)
14. Vaillant, R., Barthe, L., Guennebaud, G., Cani, M.-P., Rohmer, D., Wyvill, B., Gourmel, O., Paulin, M.: Implicit skinning: real-time skin deformation with contact modeling. ACM Trans. Graph. 32(4), 125 (2013)
15. Kavan, L., Sorkine, O.: Elasticity-inspired deformers for character articulation. ACM Trans. Graph. 31(6), 196 (2012)
16. Mohr, A., Gleicher, M.: Building efficient, accurate character skins from examples. ACM Trans. Graph. 22(3), 562–568 (2003)

Comparison of CFD-Based and Bernoulli-Based Pressure Drop Estimates across the Aortic Valve Enabled by Shape-Constrained Deformable Segmentation of Cardiac CT Images

Jochen Peters[1], Angela Lungu[2], Frank M. Weber[1],
Irina Waechter-Stehle[1], D. Rodney Hose[2], and Juergen Weese[1]

[1] Philips Research, Hamburg, Germany
[2] The University of Sheffield, U.K.

Abstract. The aortic valve area (AVA) and the pressure drop (PD) across the aortic valve are important quantities for characterizing an aortic valve stenosis. Using the Bernoulli equation and mass conservation, a relation between both quantities can be derived. We developed a simulation pipeline to assess the accuracy of this relation for realistic patient anatomies and blood flow rates. The key element of the pipeline is a shape-constrained deformable model (SCDM) for the segmentation of the aortic valve, the ascending aorta and the left ventricle over the cardiac cycle in cardiac CT images. Efficient segmentation enabled application of the simulation pipeline to cardiac CT image sequences of 22 patients. Planimetric AVA and Bernoulli-based PD estimates were computed from the same segmentation results. The resulting PD estimates show a high correlation ($R = 0.97$), but Bernoulli-based PD results are on average 25% smaller than the CFD-based results. The results contribute to a better understanding and interpretation of clinically used quantities such as the AVA and the PD.

1 Introduction

In patients with age over 65, aortic valve stenosis is most often caused by calcification that prevents proper valve opening and closing. The aortic valve opening area (AVA) provides geometric information on the severity of aortic valve stenosis while the pressure drop (PD) describes to what extent improper valve opening impairs blood flow. Using the Bernoulli equation and mass conservation, frequently used relations between AVA and PD can be derived [6].

We developed a simulation pipeline to assess the accuracy of the Bernoulli-based PD estimate for realistic patient anatomies. Its key element is a segmentation method of the heart and the aortic valve (AV) anatomy in CT image sequences. The segmentation result defines the simulation domain for a steady-state simulation of the blood flow through the AV. The left ventricular (LV) volume change defines the (maximum) flow rate. While previous work focused on comprehensive FSI simulations [1,2,9,12,13] that have mostly been applied to

F. Bello and S. Cotin (Eds.): ISBMS 2014, LNCS 8789, pp. 211–219, 2014.

very few patient cases, efficient segmentation enables application of our simulation pipeline to a considerably number of cases.

Only recently, suitable segmentation methods have been proposed. In particular, an approach to segment the aortic and mitral valve in CT and ultrasound sequences [8] has been proposed and extended for segmenting the complete valvular heart apparatus in CT image sequences [7]. Furthermore, the shape-constrained deformable model (SCDM) framework [3,4] was used to segment the LV, closed AV and aorta in cardiac CT images [14] to support quantification of AV structures in patients considered for transcatheter aortic valve interventions (TAVI).

We build our work on the SCDM-based framework [3,4]. In particular, we designed a detailed aortic valve mesh model with structured meshes and double-layered leaflets (Sec. 2). Specific model structures that stabilize model adaptation in deformable adaptation were added. Boundary detection was trained using a variant of Simulated Search that penalizes undesired boundaries [11]. Sec. 3 explains the Bernoulli- and CFD-based assessment of the PD and includes results for 22 patient cases. Our conclusions are summarized in Sec. 4.

2   Aortic Valve Segmentation

2.1   Shape-Constrained Deformable Models

Within the framework [3,4], a mesh model of the target anatomy with V vertices $m_1, \ldots, m_V$ and T triangles is adapted to an image. First, the anatomical structure is detected using the Generalized Hough Transformation (GHT). Afterwards, parametric model adaptation is performed. For that purpose, the model is transformed according to $\mathcal{T}(q)[m_k]$ with q describing the transformation parameters. Boundary points x_i^t are detected along profiles parallel to the triangle normals using individually trained boundary detectors for each triangle i, and the transformation parameters q are updated by minimizing the external energy

$$E_{\text{ext}} = \sum_{i=1}^{T} \tilde{w}_i \left(\frac{\nabla I(x_i^t)}{\| \nabla I(x_i^t) \|} (x_i^t - c_i(q)) \right)^2, \tag{1}$$

where I denotes the image and the weights $\tilde{w}_i$ are derived during boundary detection. The triangle centers $c_i(q) = \frac{1}{3} \sum_{j=1}^{3} \mathcal{T}(q)[m_{k(i,j)}]$ are calculated from the vertices $\{k(i,1), k(i,2), k(i,3)\}$ forming the triangle i. Boundary detection and refinement of the parameters q are iterated several times.

Parametric adaptation may be done in several stages. Initially, a similarity transformation $\mathcal{T}_{\text{sim}}$ may be used, and adaptation may be further refined using a multi-linear transformation

$$\mathcal{T}_{\text{multi-linear}}(q)[m_i] = \sum_{k=1}^{K} w_{i,k} \cdot \mathcal{T}_k(q_k)[m_i], \tag{2}$$

where the weights $w_{i,k}$ are used for a fuzzy subdivision of the model into K parts. In a final stage SCDM adaptation is performed by iterating boundary detection and mesh deformation. Mesh deformation optimizes the energy $E = E_{\text{ext}} + \alpha E_{\text{int}}$ composed of the external energy of eq. (1) and the internal energy

$$E_{\text{int}} = \sum_{i=1}^{V} \sum_{j \in N_i} \sum_{k=1}^{K} w_{i,k} \left(\boldsymbol{x}_i - \boldsymbol{x}_j - \mathcal{T}_k(\boldsymbol{q}_k)[\boldsymbol{m}_i - \boldsymbol{m}_j] \right)^2 . \tag{3}$$

The vertex coordinates $\boldsymbol{x}_i$ are optimized together with the transformation parameters $\boldsymbol{q}_k$. In the external energy of eq. (1), the transformed model triangle centers $\boldsymbol{c}_i(\boldsymbol{q})$ are replaced by the triangle centers of the deforming mesh. N_i is the set of neighbors of vertex i. The weight α controls to what extent deviations between adapted mesh and reference shape are penalized. In addition, the framework can activate or freeze mesh parts and use multi-resolution models.

2.2 Aortic Valve Model

To build a structured mesh model of the AV, we defined a double layered "wedge" (Fig. 1a) representing a single aortic valve leaflet. Three wedges represent the aortic valve leaflets when the valve is closed. They have been complemented by a tubular structure in direction of the ascending aorta and a tubular structure in direction of the left ventricular outflow tract (LVOT) (Fig. 1b).

The model also includes connections between the layers building an aortic valve leaflet. These connections contribute to the internal energy of eq. (3). They help to maintain the thickness of the valve leaflets and to avoid intersections of both layers. In addition, the model contains a set C_{coapt} of connections between the vertices at the coaptation edges of the valve leaflets. For a closed valve,

$$E_{\text{close}} = \sum_{(i,j) \in C_{\text{coapt}}} \left(\boldsymbol{x}_i - \boldsymbol{x}_j \right)^2 . \tag{4}$$

can be added to the energy to prevent overlapping aortic valve leaflets.

The model was semi-automatically adapted to 147 contrasted cardiac CT scans from 57 patients using the previously described bootstrap approach [3]. The CT images have an in-plane resolution of 0.30×0.30 - $0.78 \times 0.78 \, \text{mm}^2$ and a slice thicknesses of 0.33 - 1.00 mm. The data covered closed and open valve states from subjects with healthy or stenosed calcified valves.

To enable AV segmentation in different opening states, we calculated a mean mesh from a set of closed valves with vertices $\bar{\boldsymbol{m}}_i^{close}$ and another mean mesh representing healthy open valves $\bar{\boldsymbol{m}}_i^{open}$. Both mean meshes were combined as

$$\boldsymbol{m}_i(p) = \frac{1}{2} \left(\bar{\boldsymbol{m}}_i^{close} + \bar{\boldsymbol{m}}_i^{open} \right) + \frac{p}{2} \left(\bar{\boldsymbol{m}}_i^{close} - \bar{\boldsymbol{m}}_i^{open} \right) \tag{5}$$

into a point distribution model (PDM). Opening and closing of the valve is described by the parameter p (Fig. 1c and d). During SCDM adaptation, deformation modeling by multi-linear transformations is combined with the PDM by inserting eq. (5) into the internal energy of eq. (3). It should be emphasized that the opening state of the segmentation is not simply a linear interpolation of open and closed valve states, because the internal energy of eq. (3) allows the SCDM-based segmentation result to deviate from the shape model.

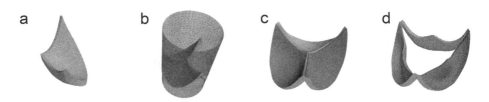

Fig. 1. Illustration of the design of the aortic valve mesh (a, b). In addition, the mean closed $(p = 1)$ and open $(p = -1)$ aortic valve leaflets are shown (c, d).

2.3 Boundary Detection

Boundary detection and training are described in [10]. Different features have been used, because the AV leaflets can be either clearly visible, noisy, or distorted by calcification. Gray values are evaluated on a hexagonal grid (19 sampling points; 1 mm distance) in planes perpendicular to the triangle normal. The image gradient $\nabla I(\boldsymbol{x})$ is approximated by the scaled difference of the average gray value on two planes on both sides of the considered boundary point. The gray values on either side of the boundary averaged over 1, 2 or 4 points along the triangle normal are used as criteria $Q_j(\boldsymbol{x})$ to suppress boundaries if the clipping criteria are outside triangle-specific acceptance intervals $[\text{Min}_{i,j}, \text{Max}_{i,j}]$. In addition, the difference of the boundary gray value and the average gray value on both sides of the boundary, the variance of several points along the triangle normal, the variance of the gray values within a hexagonal grid, and gradient values on either side of the tested boundary position are considered as criteria $Q_j(\boldsymbol{x})$.

All quantities were evaluated at the reference boundary of the 147 training images, and a large set of boundary detector candidates was generated using different parameters for clustering and combinations of criteria $Q_j(\boldsymbol{x})$. To select a boundary detector, a triangle is displaced, boundary detection is performed using a specific detector and the Euclidean boundary detection error is recorded [10,11]. It is also recorded whether a boundary is detected in an undesired region, in particular, if a boundary for the AV leaflet triangles on the side of the aortic bulbus was detected on the side of the LVOT and vice versa. Simulation of boundary detection is done for multiple displacements and all training images resulting in an average detection error $d(F, i)$ and a fraction $p(F, i)$ of undesired detections. After simulating boundary detection for all candidate functions, the detector F minimizing

$$d_{\text{penalized}}(F, i) = (1 - \beta) \cdot d(F, i) + \beta \cdot S \cdot p(F, i) \qquad (6)$$

is selected for triangle i. The parameter $S = 10\,\text{mm}$ converts counts into millimeters. The weight β controls to what extent undesired detections are penalized.

Training was done in two configurations. Coarse boundary detection was done along profiles parallel to the triangle normal with 1 mm point spacing, training used 263513 boundary detector candidates, 147 displacements, and $\beta = 0.15$. For fine boundary detection with 0.5 mm point spacing, training used 504025 candidates, 63 displacements and no penalty $(\beta = 0)$.

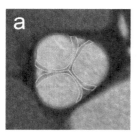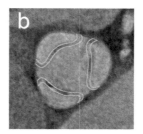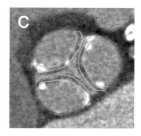

Fig. 2. Cross-sections through the AV with segmentation results of a closed valve (a), an open valve (b) and a stenosed open valve with calcification (c)

2.4   Model Adaptation and Results

The AV model was integrated into a heart model with great vessels [4] and the boundary detection functions of [11] were mapped onto the LV epi- and endocardium. Model adaptation followed the processing chain of [4]. The initial adaptation steps are performed with a half open AV ($p = 0$). During final SCDM adaptation, the AV leaflets are adapted. In a further step, the AV and LV segmentation are refined using fine boundary detection (see Sec. 2.3 for the AV and [11] for the LV). If a closed valve is detected, the energy term of eq. (4) is added during the final iteration.

The segmentation accuracy was quantified using the constrained symmetric surface-to-surface error that restricts search of the closest mesh point to a geodesic neighboorhood around the corresponding mesh point. Using a geodesic neighboorhood of 10 mm radius, a mean error of 0.6–0.8 mm has been published for the LV and the aorta [11,4]. For the AV leaflets, a geodesic radius of 5 mm was used to account for the smaller size of these structures. Evaluation was performed on 143 of the 147 training images (segmentation failed in 4 cases) by comparing segmentation result and reference annotation. Visual inspection (see Fig. 2) and a mean error of 0.47 mm provide confidence that realistic patient anatomies have been obtained.

3   Pressure Drop across the Aortic Valve

3.1   Bernoulli Equation and AVA

Fig. 3 illustrates blood flow through the AV [6]. Blood passes the LVOT with cross-sectional area A_{LVOT} and flow rate Q, traverses the AV with an opening area A_{AV}, passes the aortic bulbus, and enters the ascending aorta (AA). The (static) pressure p_{LVOT} drops while passing the AV to a minimum of p_{AV} and increases again in the AA. Using the Bernoulli equation, the pressure drop $\Delta p = p_{LVOT} - p_{AV}$ can be expressed in terms of the areas A_{LVOT} and A_{AVA}, the blood density ρ and the flow rate Q:

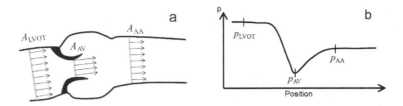

Fig. 3. 2D sketch illustrating blood flow through the AV (a) and the (static) blood pressure from LVOT to the ascending aorta (AA) (b)

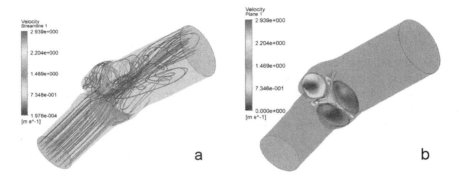

Fig. 4. Example of the model geometry used for CFD simulations. One model is displayed together with the velocity streamlines passing through the aortic valve (a) and the other model with the velocity distribution on an aortic valve plane (b).

$$\Delta p = \frac{1}{2}\rho Q^2 \left(\left(\frac{1}{A_{\text{AV}}} \right)^2 - \left(\frac{1}{A_{\text{LVOT}}} \right)^2 \right). \tag{7}$$

This equation allows to estimate the PD from geometric quantities. The flow rate Q has been computed from the LV volume change and the heart beats per minute. The cross-sectional area A_{LVOT} can easily be obtained from the segmentation result. Computation of the AVA is more complicated. In clinical studies, planimetry is used (see e.g. [5]), and an analogous approach has been implemented using the segmentation results. Starting at the aortic annulus plane, cross-sections are generated and the orifice area is measured. The plane is shifted towards the AA until the valve leaflets stop to build a contiguous opening, and the minimum orifice area is selected as AVA measurement A_{AV}.

3.2 CFD Simulations

The anatomy of the LVOT, open AV and aortic sinuses has been used as simulation domain. Inlet and outlet tubes with a length given by three times the radius have been added (see Fig. 4 for an example). The resolution of the surface mesh was increased and the surface was smoothed during this process. Volumetric meshing was performed using ICEM v14.0 and mesh sensitivity tests were carried out. For each individual case, a RMS (root mean squared) residual of pressure and momentum smaller than 10^{-5} has been chosen as a convergence criterion. Each resulting model consisted of approximately 2 million elements, typically one third of which were pentahedral elements inflated from the wall and two thirds tetrahedral elements in the core of the flow domain.

For the CFD simulations, the blood was assumed to be an incompressible Newtonian fluid with density of $1056\,kg/m^3$ and viscosity of $0.004\,Pas$. The Navier Stokes equations were solved using ANSYS-CFX v14.0, with a shear stress transport turbulence model. Two simulation results are shown in Fig. 4. The Reynolds number in the throat of the valve, based on equivalent diameter, varied between 3400 and 11100, justifying the use of a turbulence model.

3.3   Data and Results

Heart and valve segmentation was applied to 22 retrospectively gated cardiac CT angiography datasets of patients with and without AV stenosis. The datasets have been acquired on Philips CT scanners and had a resolution of $0.31-0.68\,mm$ in-plane and a slice thickness of $0.34-0.70\,mm$. The heart rate was between 51 and 98 BPM for the different patients and reconstructions at $0\%, 10\%, \ldots, 90\%$ of the heart cycle were available. Fig. 5a shows a plot of LV volume and AVA over the heart cycle for a patient with aortic valve stenosis. The valve opens when the LV starts to contract and closes when the LV relaxes again.

The PD was evaluated at the heart phase with maximum AVA. If this phase was ambiguous, the heart phase with the larger LV contraction and volume flow Q (estimated from LV volumes at heart phase $\pm 10\%$) was taken. In addition, the cross-sectional area A_{LVOT} was determined. Fig. 5b shows that the Bernoulli-based and CFD-based PD estimates are strongly correlated ($R^2 = 0.97$), but Bernoulli-based results are about 25% smaller than CFD-based results.

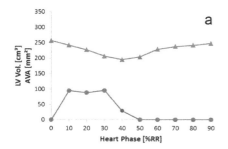 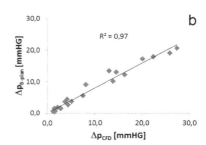

Fig. 5. Plot of the LV Volume and AVA over the heart cycle for a patient case (a) and plot of the CFD-based PD vs. the Bernoulli-based PD (b)

4   Discussion and Conclusions

Bernoulli-based and CFD-based estimates of the PD across the AV have been compared for 22 realistic patient anatomies and blood flow rates. This comparison is enabled by the automatic segmentation of the aortic valve and heart anatomy over the cardiac cycle in cardiac CT images. The method extends the SCDM framework [3,4] with new features for boundary detection and connections between surfaces that stabilize segmentation. In combination with a structured aortic valve model and parametrization of valve opening by a PDM, the leaflets can be segmented with an accuracy of 0.47 mm.

A comparison of Bernoulli-based and CFD-based PD estimates shows a high correlation ($R^2 = 0.97$), but Bernoulli-based results are about 25% smaller. The use of steady-state blood flow simulations on a fixed domain and computation of the Bernoulli-based PD estimate using the same segmentation result and flow rate limits possible origins of the deviation and excludes, for instance, segmentation errors.

A possible explanation of the deviations relates to the difference between geometric AVA given by the geometric area of the valve orifice and the effective AVA given as the minimal cross-sectional area of the downstream jet when computing the Bernoulli-based PD estimate. For an analytical stenosis model, the effective AVA can, in theory, be 40% smaller than the geometric AVA [6]. Furthermore, the complex 3D geometry of the AV may not sufficiently well be characterized by the geometric AVA derived from planimetry.

The results contribute to a better understanding of clinically used quantities such as AVA and PD. The simulation pipeline could also be used as a starting point of a CT-based method to assess the PD complementary to established approaches such as echocardiography or invasive pressure measurements.

Acknowledgment. This work was supported by NRW, Germany and the European Union ("NRW Heart valve initiative", Med in.NRW 005-GW01-235C). A. Lungu is funded by the UK Engineering & Physical Sciences Research Council.

References

1. Carmody, C.J., Burriesci, G., Howard, I.C., Patterson, E.A.: An approach to the simulation of fluid-structure interaction in the aortic valve. J. Biomech. 39(1), 158–169 (2006)
2. Hart, J.D., Peters, G.W.M., Schreurs, P.J.G., Baaijens, F.P.T.: A three-dimensional computational analysis of fluid structure interaction in the aortic valve. J. Biomech. 36(1), 103 (2003)
3. Ecabert, O., Peters, J., Schramm, H., Lorenz, C., von Berg, J., Walker, M.J., Vembar, M., Olszewski, M.E., Subramanyan, K., Lavi, G., Weese, J.: Automatic model-based segmentation of the heart in CT images. IEEE Trans. Med. Imaging 27(9), 1189–1201 (2008)

4. Ecabert, O., Peters, J., Walker, M.J., Ivanc, T., Lorenz, C., von Berg, J., Lessick, J., Vembar, M., Weese, J.: Segmentation of the heart and great vessels in CT images using a model-based adaptation engine. Med. Image Anal. 15(6), 863–876 (2011)
5. Feuchtner, G.M., Dichtl, W., Friedrich, G.J., Frick, M., Alber, H., Schachner, T., Bonatti, J., Mallouhi, A., Frede, T., Pachinger, O., zur Nedden, D., Müller, S.: Multislice computed tomography for detection of patients with aortic valve stenosis and quantification of severity. J. Am. Coll. Cardiol. 47(7), 1410–1417 (2006)
6. Garcia, D., Kadem, L.: What do you mean by aortic valve area: Geometric orifice area, effective orifice area, or gorlin area? J. Heart Valve Dis. 15(5), 601–608 (2006)
7. Grbic, S., Ionasec, R., Vitanovski, D., Voigt, I., Wang, Y., Georgescu, B., Navab, N., Comaniciu, D.: Complete valvular heart apparatus model from 4D cardiac CT. Med. Image Anal. 16(5), 1003–1014 (2012)
8. Ionasec, R.I., Voigt, I., Georgescu, B., Wang, Y., Houle, H., Vega-Higuera, F., Navab, N., Comaniciu, D.: Patient-specific modeling and quantification of the aortic and mitral valves from 4-D cardiac CT and TEE. IEEE Trans. Med. Imaging 29(9), 1636–1651 (2010)
9. Mihalef, V., Ionasec, R.I., Sharma, P., Georgescu, B., Voigt, I., Suehling, M., Comaniciu, D.: Patient-specific modelling of whole heart anatomy, dynamics and haemodynamics from four-dimensional cardiac CT images. Interface Focus 1(3), 286–296 (2011)
10. Peters, J., Ecabert, O., Meyer, C., Kneser, R., Weese, J.: Optimizing boundary detection via Simulated Search with applications to multi-modal heart segmentation. Med. Image Anal. 14(1), 70–84 (2010)
11. Peters, J., Lessick, J., Kneser, R., Wächter, I., Vembar, M., Ecabert, O., Weese, J.: Accurate segmentation of the left ventricle in computed tomography images for local wall thickness assessment. In: Jiang, T., Navab, N., Pluim, J.P.W., Viergever, M.A. (eds.) MICCAI 2010, Part I. LNCS, vol. 6361, pp. 400–408. Springer, Heidelberg (2010)
12. Shadden, S.C., Astorino, M., Gerbeau, J.F.: Computational analysis of an aortic valve jet with lagrangian coherent structures. Chaos 20(1), 017512 (2010)
13. Votta, E., Le, T.B., Stevanella, M., Fusini, L., Caiani, E.G., Redaelli, A., Sotiropoulos, F.: Toward patient-specific simulations of cardiac valves: State-of-the-art and future directions. J. Biomech. 46(2), 217–228 (2013)
14. Waechter, I., Kneser, R., Korosoglou, G., Peters, J., Bakker, N.H., Boomen, R.v.d., Weese, J.: Patient specific models for planning and guidance of minimally invasive aortic valve implantation. In: Jiang, T., Navab, N., Pluim, J.P.W., Viergever, M.A. (eds.) MICCAI 2010, Part I. LNCS, vol. 6361, pp. 526–533. Springer, Heidelberg (2010)

FE Simulation for the Understanding
of the Median Cystocele Prolapse Occurrence

Olivier Mayeur[1,2], Gery Lamblin[1,4], Pauline Lecomte-Grosbras[1,3], Mathias Brieu[1,3],
Chrystele Rubod[1,5], and Michel Cosson[1,5].

[1] Laboratoire de Mécanique de Lille, CNRS UMR-8107, France
[2] Université de Lille 2, Droit et Santé, France
[3] Ecole Centrale de Lille, France
[4] Département de Chirurgie Gynécologique,
Hôpital Femme Mère Enfant, CHU de Lyon, France
[5] Clinique de Chirurgie Gynécologique,
Hôpital Jeanne de Flandre, CHRU de Lille, France
Olivier.mayeur@univ-lille2.fr

Abstract. Female pelvic organ prolapse is a complex mechanism combining the mechanical behavior of the tissues involved and their geometry defects. The developed approach consists in generating a parametric FE model of the whole pelvic system to analyze the influence of this material and geometric combination on median cystocele prolapse occurrence. In accordance with epidemiological and anatomical literature, the results of the numerical approach proposed show that the geometrical aspects have a stronger influence than material properties. The fascia between the bladder and vagina and paravaginal ligaments are the most important anatomical structures inducing the amplitude of cystocele prolapse. This FE model has also allowed studying the coupled effect, showing a significant influence of the fascia size. The study allows highlighting the origins of the median cystocele prolapse and responds to this major issue of mobility occurrence.

Keywords: Pelvic floor, Genital prolapsed, FE modeling.

1   Introduction

Pelvic organ prolapse (POP) is a woman mobility disorder of the pelvic system organs. This hypermobility represents a major problem in our present-day society (60% of women more than 60 years old [1]) and is due to anatomical and mechanical dysfunction of the pelvic system. Aging of the female population involves growing demand for surgery to improve quality of life. The main cause of failure lies in a complex physiopathology, linked to soft tissues, fasciae and ligaments. This organ support system is misunderstood or poorly described. During the past decades, medical and engineering sciences have been targeting together a better understanding of the pelvic mobility. Research then focuses on interpreting these mechanisms which combine the mechanical properties of the tissues involved and their geometry defects.

F. Bello and S. Cotin (Eds.): ISBMS 2014, LNCS 8789, pp. 220–227, 2014.
© Springer International Publishing Switzerland 2014

The most common form of POP is the cystocele, which occurs mostly when the bladder protrudes into the anterior vaginal wall [2]. To sum up, there are three types of cystocele (median, lateral and high) which may occur in clinical situations either separately or combined. The study focuses on median cystocele prolapse (MCP) which occurs the most. The literature review reports two principal theories inducing the amplitude of cystocele. The first theory shows a link between organ mobility and the fascia between bladder and vagina (BVF) [3,4]. Adding to this study, Chen et al. has shown that MCP is also sensitive to the paravaginal support defects [2], [5,6]. To sum-up, the theory #1 (Petros) pretends that the defect of BVF is the most important cause of MCP and theory #2 (DeLancey) assess that paravaginal ligaments must be added to induce cystocele. Concerning the material properties, significant work on characterization of the mechanical behavior has been undertaken via destructive experimental tests [7,8]. All this research work enables us to map the mechanical properties of tissues and underline the major elements of pelvic floor disorder, while we still do not understand their respective roles. Other studies deal with the evaluation of the mechanical properties of tissues depending on age or pathology [7], [9,10,11].

Yet the multifactorial character of this complex pathology makes it difficult to understand the origins of hypermobility. The Finite Element (FE) method seems to be an adequate technique to understand the pelvic mobility and the role of its related support system.

2   Material and Method

Numerical simulations need accurate information about the geometrical and material properties. The generation of a FE model follows a standard guideline relating to a 3D model reconstruction, material definition, meshing and loading. Special attention is paid to the creation of the 3D model thanks to software and appropriate personal developments. The properties of materials are based on previous results already published in literature [11]. All these data allows us to generate a parametric model of the pelvic system, exploitable in different configurations to analyze the influence of each anatomical structure on bladder displacement.

2.1   Reconstruction of the Pelvic Model

Magnetic Resonance Image (MRI) data, obtained with a 3T MRI, is used to define the numerical geometry of the pelvic system. The data information is represented thanks to 3 sequences of 2D images obtained on the axial, coronal and sagittal incidences (Resolution 512x512, Pixel size: 0.586 mm). To generate a 3D representation of the pelvic system, the contour of each organ is defined semi-manually. This preliminary stage is made on Avizo software. As the MRI is based on the magnetic resonance of hydrogen molecule, it helps to visualize anatomical structures that contain water such as soft tissues or bladder. Vaginal Gel Contrast is injected into the vagina and rectum before the scan in order to improve the contrast. Threshold value can be easily defined on the image with this protocol. Thanks to this contour selection and slice superimposition, vagina (VA), rectum (RE), bladder (BL) and uterus (UT) are represented in 3D model at STL format (fig.1).

However, this study focuses on the suspension system. MRI technique does not allow viewing the ligaments due to their thinness and the low MRI resolution. This same constraint limits the definition of the pelvic floor. Another strategy has therefore been defined to represent the entire organ support mechanism, in collaboration with surgeons and anatomists, whose skills are more relevant to assess the incidence of such thin structures. Part of this anatomic structure is then generated on this sequence of slices but most of the design is achieved on the Computer Aided-Design (CAD) model.

2.2   Modeling of Anatomical Structures

The employed method consists in generating manually the surface of each organ. Each surface is defined to generate a parametric mesh in compliance with the organ geometry. The model is controlled by B-spline curves linked to the cloud of dots. This stage is conducted with special care to ensure that surfaces are compatible with well-ordered FE meshing. All organs are partitioned into regions to use quad or hexahedral elements in the following steps. The construction lines are designed to provide structured mesh patterns. This strategy is employed to help set the parameters of the FE model. Since the study deals not only with the change in the material properties but also with geometrical modifications resulting in pulled fasciae and ligaments, CAD configuration enables FE meshing to be updated automatically. The collected data will make it possible to analyze the influence of each structure on the behavior of the pelvic system. The alteration of geometries will be performed numerically on the basic model obtained from the initial reconstruction.

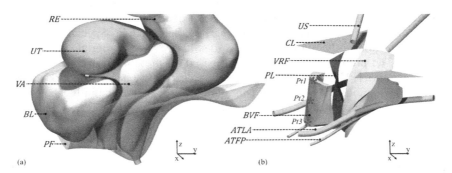

Fig. 1. Model development. (a) 3D surface-rendered model of organs. (b) 3D surface-rendered representation of anatomical support structures. RE: rectum; UT: uterus; VA: vagina; BL: bladder; PF: pelvic floor; US: uterosacral ligaments; CL: cardinal ligaments; VRF: vagina/rectum fascia; PL: paravaginal ligaments; BVF: bladder/vagina fascia; ATLA: arcus tendineus levator ani; ATFP: arcus tendineus fascias pelvis.

The bladder surface helps to simulate an incompressible fluid inside the bladder and provide a better representation of organ mobility. The pelvic floor (fig.1a) corresponds to a support sustaining organs and is rigidified by muscles and ligament structures corresponding to the Arcus Tendineus Fascias Pelvis (ATFP) and the Arcus Tendineus Levator Ani (ATLA). The other ligament structures (fig.1b) to be modeled are the Uterosacral ligament (US) maintaining the vaginal apical in position and the Cardinal Ligament (CL) between anterior cervical ring (cervix and vagina) and around the origin of the internal iliac artery (pelvic sidewall). The Paravaginal Ligaments (PL), linked to the pelvic sidewall, are the third structure supporting the vagina and are represented on both sides of it. In accordance with the theory #2, this ligament is highly influencing the MCP and will be the first parametric structure used in the FE simulation. Its geometry was designed to make it variable by increasing its length on the CAD model and updating the FE model. Its Young's modulus is variable in the rest of the study as well. The last step is to define the fasciae between bladder and vagina (BVF) and between vagina and rectum (VRF). Those structures are represented by volumes in order to generate regular hexahedral elements during the meshing method. Since the BVF is one of the structures responsible for the MCP in both theories [3,4], its FE model will take into account 2 input parameters corresponding to the material (Young's modulus) and geometrical properties.

Experimental results in literature show an increased stiffness with age or pathology [9,10,11]. Furthermore, cyclic tests show irreversible residual strains [8]. This experimental observation leads to the following interpretation: there is a geometric evolution of ligaments as their length increased under the stress effects of the support system during life. On the one hand, PL dimension may vary to take into account aging and/or pathology. On the other hand, since BVF is a constraint structure between two surfaces, the scale at which this length variation occurs is not accessible to our simulation tool. Thus we propose to take into account this geometric evolution by reducing the Young's modulus at the origin, thus keeping BVF size constant.

2.3   Properties and Data Processing

The refined model was imported in ABAQUS (Version 6.11-2) to create the FE mesh. Tab.1 sums up the different elements employed in simulation, the thicknesses of shell elements and the number of elements for each anatomical structure. In order to understand the mobility of the pelvic system, the first model generated corresponds to a healthy patient whose mechanical properties are characteristic of the values of a young woman [8], [11]. Mechanical and geometrical properties are then altered to confront the 2 theories. The material properties rely on a previous study [11] which has the advantage of providing results concerning young (Y) as well as old (O) women (tab.1). Experimental tests [11] show an increase of the tangent modulus at zero strain, equivalent to Young's modulus. For instance, for paravaginal ligaments it increases from 2.22 MPa to 2.52 MPa (O/Y ratio: 1.14). The study did not take into account the difference between old and young anterior vaginal walls. However, the bladder and the vagina present an augmentation of material properties with 2.25 and 3.55 O/Y ratio, respectively. As the BVF fascia is located between these two organs, we assume that the Young's modulus is increasing from 0.03 MPa to 0.09 MPa. Another study has compared pathologic with non-pathologic vaginal tissues [9] and has also noticed increased stiffness with pathology.

The strategy used consists in analyzing bladder displacements taking into account this growing stiffness with age. As age is an important factor affecting pelvic mobility, a preliminary study will help to understand the phenomena induced by this alteration and conclude how they are influencing the POP. The purpose of the following study is to analyze the sensitiveness of each parameter according to the two structures under study (BVF and PL). As this analysis of sensitiveness involves many parameters, we have decided to study these two anatomical areas separately. The advantage of this approach is to find how each modification affects mobility.

Table 1. Mechanical properties of pelvic soft tissues of young and old women and FE model data (Element type, thickness and number)

	E (young) MPa	E (old) MPa	O/Y ratio	Elt type	Thickness mm	Elt Nb
Rectum*	0,54	2,1	3,89	Quad	3	3.7k
Vagina*	0,66	2,34	3,55	Quad	3	2.4k
Bladder*	0,24	0,54	2,25	Quad	2	3.6k
Pelvic floor	0,03	na	na	Quad	2	3.5k
Uterosacral lgt*	0,78	4,98	6,38	Beam	na	0.03k
Cardinal lgt*	1,32	4,2	3,18	Quad	1	2.7k
Paravaginal lgt*	2,22	2,52	1,14	Quad	1	2.1k
ATLA and ATFT	0.78	na	na	Quad	2	1k
BVF	0.03	na	3	Hexa	na	7k
VRF	0.04	na	na	Hexa	na	8k

Young's modulus [11]

The analysis of the mobility of the bladder is based on the displacement in our coordinate system (x: sagittal axis; y: coronal axis; z: transverse axis) of three characteristic points of the bladder. These points are located in the upper (pt1), middle (pt2) and bottom (pt3) zones of the bladder and are taken in a sagittal plane in order to analyze the movements of the bladder to the vagina (fig.1b). This selection was chosen as the most appropriate displacement analysis to describe the mobility of a MCP. The results of the FE simulations are given for a fixed load, driven by a constant effort oriented at 45 degrees, corresponding to an intra-abdominal pressure [8], [12]. The study takes into account the small perturbations theory and focuses on the movement at the beginning of simulation to ensure numerical stability. Since the movements of the bladder are proportional to the intra-abdominal pressure, we chose not to include this parameter in our study to reduce the impact of this factor on the results. We are conscious that this study does not represent the full kinematics of MCP but it still shows the influence of each structure on the mobility of the pelvic system.

The input parameters may be summarized as 4 different configurations:

- Config1: evolution of BVF Young's modulus between 0.03 MPa and 0.09 MPa.
- Config2: evolution of PL Young's modulus between 2.22 MPa and 2.52 MPa.
- Config3: evolution of BVF geometry.
- Config4: evolution of PL geometry.

3   Results

The first numerical simulation (without defect of BVF and PL) reveals an initial movement of the pelvic organs system. This numerical simulation is then studied (fig.2) in extreme cases corresponding to pathological or aged tissues whose modulus is higher for BVF and PL (config1 and 2).

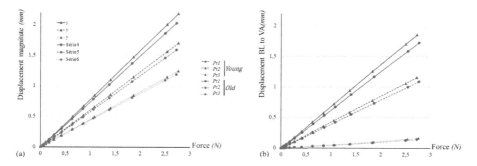

Fig. 2. Force displacement curves of the 3 points for Young (initial) and Old (config.1 and 2). (a) Displacement magnitude. (b) Displacement of the bladder to the vagina.

The confrontation between models with the mechanical properties of young and old tissues showed a decrease in bladder movement. For the equivalent force, bladder moves less in the old tissues configuration than the young one. The magnitude of this reduction is about 8% on top and middle zones of the anterior vagina wall (fig2.a). The gap between the 2 configurations is narrower in the bottom zone (pt3; 4%). The displacement of the bladder to the vagina (fig2.b) shows a similar tendency. Displacement is almost equivalent with and without changes in the lower part (pt3). These results show that hypermobility is not a phenomenon related to the material evolution.

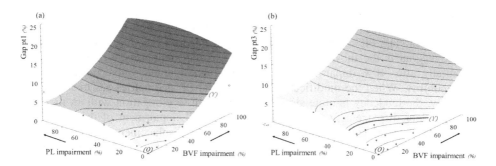

Fig. 3. Surface representation of the gap between old (O) tissues properties and old with geometrical impairments, (a) on the top of the bladder, pt1. (b) on the bottom of the bladder, pt3. The red lines correspond to the initial simulation with the young (Y) tissues properties.

The pathology occurrence is rather geometric, favoring evolution of structures following an increase of BVF (config.3) and PL (config.4). The geometric changes correspond to 10, 20, 40 and 60% elongation for PL and equivalent evolution between 10 and 80% for the BVF. The comparison takes into account the properties of the aged tissues. The difference for each point is then calculated from this simulation. The BVF and PL impairments are then coupled in order to generate a 3D representation of this gap and analyze the influence of impairment proportion. Twenty-five FE simulations are used to plot surfaces corresponding to the gap on the top (fig3.a) and bottom (fig3.b) of bladder. The results observed in this comparison (fig.3) show an increase of the gap between the aged tissues model and aged tissues with geometric evolution corresponding to an increase of bladder displacement. With a geometric impairment of BVF (config.3), the difference is significant on the magnitude of displacement for each point studied. There is a strong increase after 40% degradation (fig.3a, BVF impairment axis). PL elongation (config.4) presents a different trend with a less pronounced difference after the 40%, considering a level near 6% after 60% gap (fig.3a, PL impairment axis). Before this value, the impairment of the 2 studied structures is comparable with less than 4% gap between with and without geometrical changes. This representation could be compared to the initial simulation with young tissues represented by red lines (fig.3).

4   Discussion and Conclusion

Thanks to a refined FE model of the woman pelvic, this study enables us to analyze the influence of each support structure on MCP. This parametric model was created with special care thanks to MRI data, CAD tools and the skills of surgeons and anatomists. This protocol helps to define the necessary starting point to ensure a representative FE simulation. This approach aims at configuring the FE model according to several parameters, not only material but also geometric.

As a first step, a preliminary study has shown that the increase of mechanical properties of pathological and aged tissues, reported in literature [11], leads to a decrease in organ mobility. Then this observation underlines the importance of taking into account the alteration of the geometry in the studies related to the mobility of the pelvic system. The FE simulation results have revealed that the fascia between bladder and vagina is the most influence structure in the median cystocele occurrence (essentially after 40% impairment). The paravaginal ligament impairment also impacts mobility. When the modifications in BVF and PL geometries are treated independently, the tendency is quite similar before this 40% elongation with an increase of the bladder displacement. After this value, a small impairment of the fascia between bladder and vagina has a higher influence on the displacement than paravaginal ligaments. This FE model has also permitted the study of the coupled effect of both structures.

Pelvic system is well known, but the causes of pelvic mobility are still poorly understood. Consequently, the challenge for surgeons is to know what happens at the beginning of cystocele and identify the influence of each structure. These causes are multifactorial and then rely on complex biomechanical knowledge. The FE simulation tool allows us to highlight the origin of this pathology and responds to this major issue of mobility occurrence. This study provides part of the answers but it would be interesting to complete it by studying with more deformation. The POP is not integrally represented with this preliminary model as the study takes into account the small perturbations theory. Since the movement is proportional to the intra-abdominal pressure, future works will take into account this modification in loading condition and hyper-elasticity.

Additional research with the same approach could also be conducted to compare other structures (ATLA, ATFP, US) and to analyze theirs influences on lateral or high cystocele. As explained previously, this protocol allows us to transform the material behavior and geometrical properties of each supporting structure. In the same way, we could plan to morph the organ geometries (vagina length, bladder size) to study their influence on the POP and finally to perform dedicated models for each patient. As the geometry has a significant effect on organ mobility, it will be interesting to generate new models from other patients in order to confirm those preliminary results.

The results lead to interesting questions concerning the geometrical impairments. The numerical observation must be compared with an anatomical exploration, as it has been shown recently that elongation of cardinal and uterosacral ligaments might be noticed [13]. It will be interesting to study the elongation of these structures during aging process and in function of life quality.

References

1. Samuelsson, E., Victor, F., Tibblin, G., Svardsudd, K.: Signs of genital prolapse in a Swedish population of women 20 to 59 years of age and possible related factors. Am. J. Obstet. Gynecol. 180, 299–305 (1999)
2. DeLancey, J.O.L.: Fascial and Muscular Abnormalities in Woman with Urethral hypermobility and Anterior Vaginal Wall Prolapse. Am. J. Obstet. Gynecol. 187, 93–98 (2002)
3. Petros, P.E., Woodman, P.J.: The Integral Theory of Continence. Int. Urogynecol. J. Pelvic Floor Dysfunct., 35–40 (2008)
4. Petros, P.: The Integral System. Cent. European J. Urol. 3, 110–119 (2011)
5. Chen, L., Ashton-Miller, J.A., DeLancey, J.O.L.: A 3D Finite Element Model of Anterior Vaginal Wall Support to Evaluate Mechanisms Underlying Cystocele Formation. J. Biomech. 42(10), 1371–1377 (2009)
6. DeLancey, J.O.L.: Anatomic aspects of vaginal eversion after hysterectomy. Am. J. Obstet Gynecol., 1717–1724 (1992)
7. Goh, J.T.: Biomechanical Properties of Prolapsed Vaginal Tissue in pre and postmenopausal Women. Int. Urogynecol. J. Pelvic Floor Dysfunct. 13, 76–79 (2002)
8. Rubod, C., Brieu, M., Cosson, M., Rivaux, G., Clay, J.C., Gabriel, B.: Biomechanical Properties of Human Pelvic Organs. J. Uro. 79(4), 1346–1354 (2012)
9. Clay, J., Rubod, C., Brieu, M., Boukerrou, M., Fasel, J., Cosson, M.: Biomechanical Properties of Prolapsed or non-Prolapsed Vaginal Tissue: Impact on Genital Prolapse Surgery. Int. Urogynecol. J. 12, 1535–1538 (2010)
10. Gilchrist, A., Gupta, A., Eberhart, R., Zimmern, P.: Biomechanical Properties of Anterior Vaginal Wall Prolapse Tissue Predict Outcome of Surgical Repair. J. Uro. 18(3), 1069–1073 (2010)
11. Chantereau, P., Brieu, M., Kammal, M., Farthmann, J., Gabriel, B., Cosson, M.: (2014), http://dx.doi.org/10.1007/s00192-014-2439-1
12. Rao, G.V., Rubod, C., Brieu, M., Bhatnagar, N., Cosson, M.: Experiments and FE modeling for the Study of Prolapsed in the Pelvic System. CMBBE 13(3), 349–357 (2010)
13. Ramanah, R., Berger, M.B., Parratte, B.M., DeLancey, J.O.: Anatomy and histology of apical support: a literature review concerning cardinal and uterosacral ligaments. Int. Urogynecol. J., 1483–1494 (2012)

The Role of Ligaments:
Patient-Specific or Scenario-Specific?

Julien Bosman[1,2], Nazim Haouchine[1,2], Jeremie Dequidt[1,2], Igor Peterlik[3,4],
Stéphane Cotin[1,2,3], and Christian Duriez[1,2]

[1] Shacra Team, INRIA, France
[2] Lille University, France
[3] IHU Strasbourg, France
[4] Institute of Computer Science, Masaryk University, Czech Republic

Abstract. In this paper, we present a preliminary study dealing with
the importance of correct modeling of connective tissues such as ligaments in laparoscopic liver surgery simulation. We show that the model
of these tissues has a significant impact on the overall results of the simulation. This is demonstrated numerically using two different scenarios
from the laparoscopic liver surgery, both resulting in important deformation of the liver: insufflation of the abdominal cavity with gas (*pneumoperitoneum*) and manipulation with the liver lobe using a surgical
instrument (*grasping pincers*). For each scenario, a series of simulations
is performed with or without modeling the deformation of the ligaments
(fixed constraints or biomechanical model with the parameter of the literature). The numerical comparison shows that modeling the ligament
deformations can be at least as important as the correct selection of the
patient-specific parameters, nevertheless this observation depends on the
simulated scenario.

1 Introduction

In the context of patient-specific simulations there is a strong need of precise
biomechanical models, which are capable to capture and predict the deformations
of the organs targeted during given procedure. The ultimate goal is to use these
models in the *pre-operative* context in order to improve procedure planning as
well as in the *per-operative* context, where model-based real-time simulation is
used as a tool for guidance directly in the operation room.

When high accuracy of the biomechanical models is required, the finite element method seems to be the best approach to capture the deformations of
tissues. This method is employed to integrate viscoelastic or hyperelastic constitutive laws over the geometric domain of an organ [3]. Intensive work has been
done on measurement and estimation of soft tissue constitutive laws [5,2], nevertheless, it is still unclear where the effort should be put to obtain *accurate* and
realistic results [4]. We propose to investigate the role of the surrounding tissues and the boundary conditions that are often under-estimated. For instance,
while the constitutive laws of the liver tissues have been well studied [8,6], to

F. Bello and S. Cotin (Eds.): ISBMS 2014, LNCS 8789, pp. 228–232, 2014.

our best knowledge, there are only few existing works focusing on the role of the boundary conditions and the anatomical structures surrounding the liver.

In this paper, we focus on the role of ligaments; more precisely, we show that the modeling of ligament deformation has a significant impact on the overall simulation results. We use two different scenarios, typically performed in the context of laparoscopic surgery, both resulting in important deformations of the liver: insufflation of the abdominal cavity with gas (pneumoperitoneum) and manipulation of the liver lobe using a surgical instrument (graspers). For each scenario, a series of simulations is performed using different methods for modeling of ligaments (fixed constraints and biomechanical model). We show that removing the ligament deformation has a significant influence on the simulation of the liver. We compare this to the influence of the constitutive law used to model the liver deformation, using numerical test based on under- or over-estimation of the material properties. Then we note that given the type of input loading used to deform the model (displacement or effort), the influence changes.

2   Methods

We perform the study in the context of laparoscopic liver surgery: the laparoscopic scene represents the right and left lobes separated on the anterior side by the falciform ligament (Fig. 1). The 3D model of the lobes is obtained from pre-operative CT-scans including the vascular network [6]. The falciform ligament is more difficult to extract from the pre-operative images, however, recent work [7] shows promising results in the transfer of the ligament positions using atlas-based techniques. Thus, we assume that the ligament position can be determined. The underlying fat supporting the liver is also modeled to simulate the surrounding connective tissues.

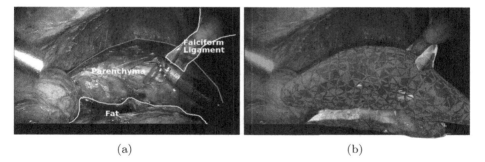

(a)   (b)

Fig. 1. Laparoscopic view of liver showing (a) contours of lobes, falciform ligament and surrounding fat, (b) superimposed FE meshes generated from pre-operative CT scans

We perform three experiments: first, the liver is deformed with a surgical instrument with known position. Second, a force similar to that imposed by the instrument is applied to the surface of the liver. Third, liver deformation due to pneumoperitoneum is simulated given the pressure applied to the patient's abdomen. The important difference between these experiments is in the type of load that creates the deformation: in the first case, a displacement is imposed on a part of the mesh (in the region of the liver that is being grasped), whereas in the second and third cases a force and a pressure (an effort) is imposed on the surface of the model. In both cases, the goal is to obtain the displacement field of the nodes of the liver model.

As the aim of this work is to study the influence of boundary conditions, the simulation involves several tests in which we combine different methods to simulate the ligaments. In each test, the liver parenchyma is simulated with the corotational formulation of the finite element method employing the linear constitutive law.

As for the ligaments, two different simulations are performed, each being based on a different hypothesis. According to the first hypothesis, the ligament is considered a full anatomical structure and simulated using linear elasticity. Alternatively, the ligaments are considered stiff enough to be modeled as fixed points on the liver surface.

3   Results

The simulations are performed using SOFA framework[1]. The implicit integration based on *backward Euler method* is employed and the linearized system of equations is solved in each time step using the *preconditioned conjugate gradients*. The volume meshes are composed of linear tetrahedral elements.

In our reference simulation, the liver and the ligaments are simulated using finite element corotational formulation. In the three experiments, we compare the results of each test to the reference simulation. The error evaluation on the displacement is done after the equilibrium is achieved using *root mean square error* (RMSE).

The tests are organized as follows: first we compare the role of the ligaments which are modeled either as fixed points or using using the corotational formulation with Young's modulus of $E_{lig} = 150$ kPa. Then, in the second and third tests, ligaments are simulated with FEM, but we under- and over-estimate the Young's modulus of the parenchyma (initially set to $E_{ref} = 27$ kPa) by the factor of two w.r.t. the reference simulation.

The grasping is simulated by prescribing a displacement on the lobe of the liver. The motion of the displaced nodes is extracted from a video of a laparoscopic surgery. The maximum displacement is 75 mm. To ensure that both prescribed displacements/efforts tests result in a similar motion, the prescribed force value is computed from the prescribed displacements scenario. We measure the average elastic force of the displaced nodes of the liver after the equilibrium. From our experiment, a traction force of 15.42 N has to be applied on the nodes to be equivalent to the prescribed displacements. The direction of the force vector is chosen to be the same as the one of the displacement vector. Displacements and

---

[1] http://www.sofa-framework.org

Table 1. Evaluation of the RMSE (in mm) for prescribed displacement (grasping) and prescribed forces (traction and pressure), compared to our reference simulation

Experiments	Grasping	Traction	Pressure
Fixed nodes	2.75	7.19	3.53
$E = 0.5 \times E_{ref}$	1.12	7.76	8.40
$E = 2 \times E_{ref}$	1.77	4.85	3.93

efforts are prescribed on the sames nodes. The pneumoperitoneum is modeled by applying a pressure of 12 mmHg on the liver surface, which is an average pressure used by surgeons during abdominal laparoscopic surgery [1].

The results presented in Tab. 1 show two specific scenarios: simulations where the deformation is induced by prescribing a displacement and those where applied forces are prescribed, each of them having their own requirements.

Considering prescribed displacements, our results show that a large difference in the elasticity parameter of the liver has a relatively small impact on the error when compared to the influence of the modeling of the ligaments. The influence of the model used for ligaments is more than twice as important as the elasticity parameter used for the parenchyma. The results demonstrate the fact that in such a scenario, the use of a physically-based deformable model is crucial.

As for the prescribed forces, the influence of variation in Young's modulus of the parenchyma is much more significant than in the case of prescribed displacements. The results emphasize the importance of using patient-specific data in such scenario. However, in this case, they also show that the influence of the model used for ligaments is comparable to the impact of the elasticity parameters.

Going further, our results also show that the difference between errors due to the ligament model and to the elasticity parameter is important when simulating the same action, in our case pulling on a liver lobe, using two different loading methods (prescribed displacement/prescribed forces).

4   Conclusion

In this paper, we demonstrated that the modeling of ligaments is at least as important as the material parameters used for parenchyma model. It highlights the need for a good modeling of the boundary conditions. Thus, considering only the patient-specific data in simulations decreases their accuracy. This highlights the fact that the accuracy of the soft tissue parametrization has two key aspects: "patient-specific" as well as "scenario-specific". Moreover, we believe that in further studies and discussions, it would be interesting to focus on influence of boundary condition modeling in terms of their correct location as well as the model chosen to simulate their inherent mechanical behaviour.

References

1. Bano, J., Hostettler, A., Nicolau, S.A., Cotin, S., Doignon, C., Wu, H.S., Huang, M.H., Soler, L., Marescaux, J.: Simulation of pneumoperitoneum for laparoscopic surgery planning. In: Ayache, N., Delingette, H., Golland, P., Mori, K. (eds.) MICCAI 2012, Part I. LNCS, vol. 7510, pp. 91–98. Springer, Heidelberg (2012)
2. Conte, C., Masson, C., Arnoux, P.J.: Inverse analysis and robustness evaluation for biological structure behaviour in FE simulation: application to the liver. Computer Methods in Biomechanics and Biomedical Engineering 15(9), 993–999 (2012)
3. Miller, K., Dane Lance, G.J., Wittek, A.: Total lagrangian explicit dynamics finite element algorithm for computing soft tissue deformation. Communications in Numerical Methods in Engineering 23(2), 121–134 (2007)
4. Miller, K., Lu, J.: On the prospect of patient-specific biomechanics without patient-specific properties of tissues. Journal of the Mechanical Behavior of Biomedical Materials 27, 154–166 (2013)
5. Payan, Y.: Soft tissue biomechanical modeling for computer assisted surgery, vol. 11. Springer (2012)
6. Peterlík, I., Duriez, C., Cotin, S.: Modeling and real-time simulation of a vascularized liver tissue. In: Ayache, N., Delingette, H., Golland, P., Mori, K. (eds.) MICCAI 2012, Part I. LNCS, vol. 7510, pp. 50–57. Springer, Heidelberg (2012)
7. Plantefève, R., Peterlik, I., Courtecuisse, H., Trivisonne, R., Radoux, J.-P., Cotin, S.: Atlas-based transfer of boundary conditions for biomechanical simulation. In: Golland, P., Hata, N., Barillot, C., Hornegger, J., Howe, R. (eds.) MICCAI 2014, Part II. LNCS, vol. 8674, pp. 33–40. Springer, Heidelberg (2014)
8. Umale, S., Chatelin, S., Bourdet, N., Deck, C., Diana, M., Dhumane, P., Soler, L., Marescaux, J., Willinger, R.: Experimental in vitro mechanical characterization of porcine glisson's capsule and hepatic veins. Journal of Biomechanics 44, 1678–1683 (2011)

Experimental Characterization and Simulation of Layer Interaction in Facial Soft Tissues

Johannes Weickenmeier[1],[*], Raphael Wu[1], Pauline Lecomte-Grosbras[2],
Jean-François Witz[2], Mathias Brieu[2], Sebastian Winklhofer[3],
Gustav Andreisek[3], and Edoardo Mazza[1],[4]

[1] Department of Mechanical and Process Engineering, ETH Zurich, Switzerland
`weickenmeier@imes.mavt.ethz.ch`
[2] Laboratoire de Mécanique de Lille, Université Lille Nord de France, France
[3] Department of Radiology, University Hospital Zurich, Switzerland
[4] Swiss Federal Laboratories for Materials Science and Technology,
EMPA Duebendorf, Switzerland

Abstract. Anatomically detailed modeling of soft tissue structures such as the forehead plays an important role in physics based simulations of facial expressions, for surgery planning, and implant design. We present ultrasound measurements of through-layer tissue deformation in different regions of the forehead. These data were used to determine the local dependence of tissue interaction properties in terms of variations in the relative deformation between individual layers. A physically based finite element model of the forehead is developed and simulations are compared with measurements in order to validate local tissue interaction properties. The model is used for simulation of forehead wrinkling during frontalis muscle contraction.

Keywords: tissue layer interaction, facial soft tissues, finite element modeling, ultrasound, wrinkles.

1 Introduction

Understanding the mechanical behavior of facial soft tissues is of great importance for many clinical applications. Physically-based deformation mechanisms describing the tissue response during muscle activation for facial expressions, wrinkle formation, and aging enhance the predictive capabilities of simulations for surgery planning, implant design, diagnosis, and for the animation industry. The experimental characterization of mechanical properties of individual tissues and their interactions, the development of corresponding constitutive model equations, and their implementation into a numerical framework for robust simulations represent key steps towards realistic simulations providing substantial insight into the complex nature of facial soft tissues.

The mechanical characterization of soft biological tissues aims at determining the highly nonlinear, anisotropic, time dependent, and often loading history

---

[*] Corresponding author.

F. Bello and S. Cotin (Eds.): ISBMS 2014, LNCS 8789, pp. 233–241, 2014.

dependent material response. Several different in- and ex-vivo measurement methods have been proposed in literature suitable for the assessment of individual tissues, tissue structures and the interaction of individual facial tissue layers. Barbarino et al. applied suction experiments to evaluate skin and deeper layers [2]. Hendriks et al. [6] present a combined suction based experimental and finite element (FE) modeling approach allowing to quantify the relative contribution of different skin layers (e.g. epidermis and dermis) to the overall tissue response. The specific method provides an estimation of material properties and a description of the connection between the epidermis and dermis layer of human skin. The related through-plane layer behavior of full-thickness skin tissue was investigated by Gerhardt et al. [5] in shear experiments. Real-time video recording captured skin layer deformations used to perform a displacement, strain, and stiffness analysis as well as the assessment of tissue layer interaction.

In-vivo quantitative visualization of tissue deformation provides essential information on the mechanical interaction between individual layers. Ultrasonography is a widely used measurement method due to its highly flexible applicability, the possibility of combining it with real-time mechanical testing, and good spatial resolution. Analysis of ultrasound images during mechanical tests provides displacement fields and corresponding strain mappings for the quantification of tissue properties as presented by Tang et al. [12] for porcine sclera; Vogt and Ermert [13] and Diridollou et al. [4] for skin tissue. Real-time ultrasound measurements allow to visualize tissue behavior due to voluntary muscle contraction in terms of tissue motion and relative tissue deformation. Such experiments are presented by Wu et al. [18] who investigated masseter muscle tissue motion and shape changes during active contraction as well as tissue interaction between muscle and neighboring tissues.

Our work aims at improving with respect to existing FE models [3,17,14] in terms of physical relevance of soft tissue geometry and mechanical response. In recent papers we have demonstrated the implementation of advanced active and passive constitutive models of face tissue [15,16], as well as the accurate representation of anatomical features, and realistic boundary conditions [3]. Our model allows to simulate facial expressions and wrinkling by incorporating experimentally quantified interaction properties of individual tissue layers. In contrast to the general assumption of tight contact between all tissue layers, there are distinct tissue interfaces that exhibit significant relative tissue movement upon shearing. The present paper describes the procedure applied to determine the mechanical properties of the interaction between specific tissue layers in different regions of the forehead by means of ultrasound-based visualization of the tissue displacement field during pull-experiments. The anatomy of the forehead displays a layered organization of tissues characterized by a distinct local variation in their interactions [8,9]. The forehead is easily accessible for ultrasound measurements and is clearly bound by skin surface and skull ensuring sufficient repeatability and high accuracy in measurement data. The experimentally observed through-layer tissue deformation and corresponding properties of layer interaction are projected onto an anatomically detailed finite element model of the forehead. This FE model serves as a benchmark for evaluating layer inter-action properties in the simulation of the pull-experiments and the formation of forehead wrinkles upon muscle contraction.

2   Experimental Characterization of Tissue Interaction

The forehead is a layered tissue structure consisting of skin, subcutaneous fat, galea aponeurotica, loose areolar tissue, muscle, and periosteum. The two layers of skin and subcutaneous fat are present across the whole forehead and are rather constant in thickness. The distribution of the other tissues varies locally, and the temporal fusion line generally divides the medial and temporal zone; it consists of stiff connective tissue that inserts onto the skull, see ultrasound images in Figure 1 and work by [9].

The frontalis muscle is the main active tissue in the forehead and it is primar-ily responsible for lifting the eyebrows during facial expressions and the resulting formation of forehead wrinkles. The muscle fibers insert into the dermis in the lower forehead enabling a maximum tissue lift upon contraction. Movement of the eyebrow is enhanced by loose areolar tissue and galea fat pads surround-ing the frontalis muscle which form a glide plane allowing for reversible relative movement between superficial skin and subcutaneous tissue and the deeper tis-sues [8,7,11].

The locally varying tissue interactions are characterized by means of ultra-sound-based visualization of the through-layer displacement field upon applica-tion of external skin displacement. The experimental setup is based on work presented by [1] and consists of a chin rest as used in ophthalmology which was modified to include a clamping device for the ultrasound probe. The ex-perimental setup was optimized for (i) reproducible alignment of the subject's forehead and the ultrasound probe for multiple consecutive measurements, (ii) an adjustable orientation of the transducer with respect to the curvature of the forehead for optimal image quality, (iii) minimal movement between subject and transducer during individual measurement sequences, and (iv) flexibility in mea-surement site and displacement magnitude. Measurements were performed on a 29 year old subject using a *GE Logiq E9* ultrasound machine with a *L8-18i-D* broad-spectrum linear transducer operating at 15 MHz with a field of view of 25mm. 6 seconds video sequences (33 fps) were recorded to capture a full loading and unloading cycle consisting of a horizontal displacement of a tape attached to the skin. Location dependence of tissue response was assessed by measuring in the medial and temporal forehead including the temporal fusion line.

The image sequences were analyzed in order to extract the deformation field within the soft tissue structure. Based on the optical flow tracking algorithm introduced by Lucas and Kanade [10], multiple regular grids were aligned tan-gentially to the surface of the skin in order to visualize the gradient of deforma-tion through the individual tissue layers. Densely connected tissue layers show a rather homogeneous displacement field across the layer boundary, while loosely connected tissues experience a pronounced gradient. Moreover, the deformation

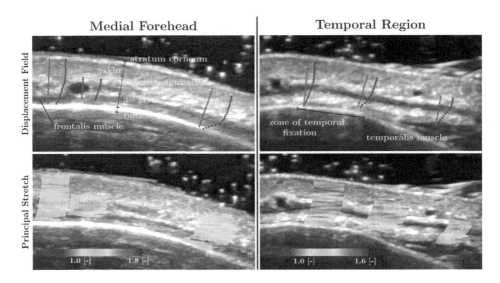

Fig. 1. Ultrasound measurements in the medial and temporal forehead. Tissue layer interaction is quantified through the evaluation of the displacement field and principal stretch in the measurement plane.

distribution provides an indirect measure of the stiffness of individual layers. The difference in the gradient between two neighboring layers depends on the stiffness ratio between both tissues. Figure 1 shows the displacement field and maximum principal stretch for a representative measurement in the medial and the temporal forehead region.

The measurement results are found to provide significant quantitative and qualitative information on the mechanical properties between individual layers and comprehensively visualize the effects of the highly differentiated forehead anatomy on the tissue response. The image resolution allows to distinguish between three characteristic main tissue layers (see Figure 1): (i) skin as the most superficial layer including the stratum corneum which appears very bright due to reflections at the boundary between gel and skin surface; (ii) SMAS or the subcutaneous tissue layer; (iii) the third layer consists of muscle fibers that are embedded in loose areolar tissue. The structure of the third layer varies significantly when moving across the forehead, i.e. the temporalis muscle lies on top of the temporal bone, the frontalis muscle covers the medial forehead, and the temporal fusion line represents the transition zone between medial and temporal forehead. The temporal fusion line is characterized by a strong interaction between all layers and provides substantial support to the whole forehead.

In general, both measurement sites show a similar behavior in terms of the deformation gradient in the two most superficial layers. Skin and subcutaneous tissue exhibit a homogeneous displacement field and a similar and almost constant deformation gradient across both layers. The difference in stiffness between skin and underlying tissue is indicated by a minor change in the slope of the displacement gradient as visible in Figure 1 at the transition from the most superficial to the second tissue layer. The third layer, however, shows a locally dependent but consistently pronounced drop in the displacement magnitude. The externally applied displacement of skin propagates through the rather stiff superficial layers all the way to the third layer. The much softer third layer of loose areolar and fat tissue exhibits significant shearing to compensate for the propagated displacement. The significant shear response of the loose areolar tissue, expressed in a large relative movement between the second layer and bone, is often associated with a gliding response of individual layers [8]. This relative movement is fully reversible upon unloading (e.g. relaxation of the muscle after a facial expression) and is apparent in the full recovery of the initial tissue state. This interaction property is evident in both, medial and temporal region. However, the measurements strongly indicate that specifically in the zone of fixation (i.e. temporal fusion line) this relative motion is fully inhibited by the strong connectivity between all layers. This is most visible in the plots of the temporal region which are characterized by a homogeneous deformation field and a very smooth gradient in comparison to the medial forehead. Finally, the principal stretch provides an additional measure to quantify the individual tissue properties.

3   Simulation of Tissue Response

An anatomically detailed finite element model of the forehead was reconstructed from MR images and consists of a multi-layered tissue structure including skin, subcutaneous tissue, a third layer with locally dependent material properties, periosteum, and temporalis muscle. The frontalis muscle is embedded in the soft areolar tissue of the third layer and inserts into subcutaneous tissue and skin close to the region of the eyebrows enabling the lift of surrounding tissue during facial expressions and wrinkle formation due to muscle contraction and tissue compression. The lower part of the face is included in this model for visualization purposes. The forehead model introduced here was generated similar to the procedure presented in [3].

Skin, subcutaneous tissue, and the third layer are tied at their individual layer surfaces to form a densely connected tissue structure. Based on the experimental observations shown in Figure 1, the interaction between the lower surface of the third layer and periosteum varies locally. In the medial forehead and superficial to the temporalis muscle, the third layer is free to slide on top of the periosteum. In proximity of the temporal fusion line, these two tissue layers are tied together, to accurately represent the dense connectivity between all tissues in the transition zone from medial to temporal forehead as described by [8]. The mechanical behavior of the active muscle and passive soft tissue layers is based on the implementation of constitutive material models as presented in [15,16]. The corresponding material parameters for facial skin were determined from suction-based in-vivo experiments on the same subject.

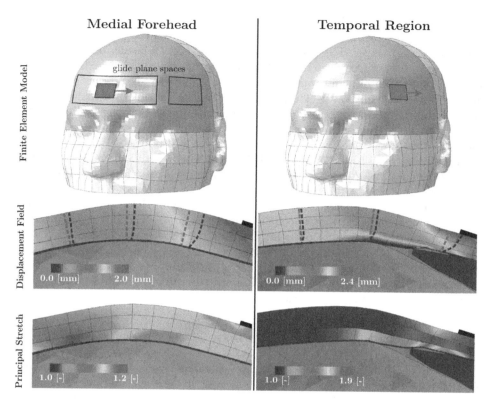

Fig. 2. FE model of pull-experiments (tape in experiments modeled by the rigid element in blue; black squares indicate regions of frictionless contact). Through-layer deformation and principal stretch fields for a sagittal cut close to the loading point.

The forehead model is used to investigate the experimentally observed tissue behavior and to validate the proposed interaction properties. The results from the numerical evaluation of the pull-experiments are shown in Figure 2 in terms of the through-layer displacement field and the corresponding principal stretch in a sagittal cut close to the loading point. A comparison of displacement magnitudes from simulation and experiment for selected points revealed an error of less than 20% which shows the significant predictive capability of the model with respect to location dependent tissue deformation and pronounced shearing. The displacement fields represent same material points at their initial position and at maximum displacement. The sliding of the third layer over periosteum causes enhanced mobility of all superficial layers which is clearly visible in the large displacement of material points across the entire thickness. Modeling the pronounced intergrowth of tissues in the zone of fixation as tightly connected layers attached to the layer of periosteum yields good agreement with the experiments showing limited displacement close to the bone and a small displacement gradient across subcutaneous tissue and skin. The connection between tissue

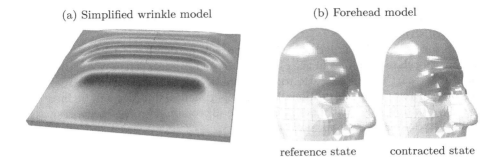

(a) Simplified wrinkle model   (b) Forehead model

reference state   contracted state

Fig. 3. (a) Deformation pattern of a simplified forehead wrinkling model (b) Simulation of frontalis muscle contraction (as in facial expressions) using the full benchmark forehead model

layers corresponds to contact conditions for which tight contact and free sliding represent limiting cases. The proposed configurations simplify the anatomical features of (i) connective collagen fibers between individual tissue layers, (ii) the embedment of frontalis muscle fibers in surrounding tissue, and (iii) the observable shearing behavior of loose areolar tissue in the forehead. However, given the predictive capabilities of the model with respect to the experimental observations, the proposed interaction properties are well justified.

A simulation of wrinkle formation is numerically challenging and requires a sophisticated numerical framework with respect to robust numerical implementations of material models and a suitable geometric discretization of the individual structures. Figure 3 shows a simplified FE model to investigate typical forehead wrinkle patterns and the simulation results based on the forehead model for the case of active frontalis muscle contraction and the resulting wrinkle formation. The model including the proposed layer interaction properties provides a suitable representation of forehead wrinkles.

4   Conclusions

An experimental setup for the visualization of full-thickness facial tissue deformation during external skin displacement and facial expressions was developed and allowed to determine location dependent properties of layer interaction in terms of inter-tissue displacement gradients. Our measurements suggest to differentiate between two different configurations: (i) a strong connection between two layers (i.e. for skin and subcutaneous tissue or for all tissues close to the temporal fusion line) and (ii) a very loose connection as in free sliding (i.e. for tissues above the eyebrow in order to enable maximum lift upon frontalis muscle contraction during facial expression).

An anatomically detailed forehead model was reconstructed from MR images and used in the simulation of pull-experiments to validate the proposed interaction properties between specific tissue layers. The physically based representation of tissue interaction properties made it possible to simulate the formation of wrinkles.

In future work, the experimental method will be employed to quantify age- and health-related changes in the observed tissue response. These measurements could form part of a diagnostic tool to quantify the impact of dermatological diseases such as fibrosis. Additionally, our experimental findings will be incorporated in the finite element model of the face [3] to simulate the tissue response during facial expression and mastication.

References

1. Barbarino, G., Jabareen, M., Mazza, E.: Experimental and numerical study of the relaxation behavior of facial soft tissue. PAMM 9, 87–90 (2009)
2. Barbarino, G., Jabareen, M., Mazza, E.: Experimental and numerical study on the mechanical behaviour of the superficial layers of the face. Skin Research and Technology 17, 434–444 (2011)
3. Barbarino, G., Jabareen, M., Trzewik, J., Nkengne, A., Stamatas, G., Mazza, E.: Development and validation of a three-dimensional finite element model of the face. Journal of Biomechanical Engineering 131(4), 041006 (11 pp.) (2009)
4. Diridollou, S., Berson, M., Vabre, V., Black, D., Karlsson, B., Auriol, F., Gregoire, J., Yvon, C., Vaillant, L., Gall, Y.: An in vivo method for measuring the mechanical properties of the skin using ultrasound. Ultrasound in Medicine & Biology 24(2), 215–224 (1998)
5. Gerhardt, L., Schmidt, J., Sanz-Herrera, J., Baaijens, F., Ansari, T., Peters, G., Oomens, C.: A novel method for visualising and quantifying through-plane skin layer deformations. Journal of the Mechanical Behavior of Biomedical Materials 14, 199–207 (2012)
6. Hendriks, F., Brokken, D., Oomens, C., Bader, D., Baaijens, F.: The relative contributions of different skin layers to the mechanical behavior of human skin in vivo using suction experiments. Medical Engineering and Physics 28(3), 259–266 (2006)
7. Hicks, D.L., Watson, D.: Soft Tissue Reconstruction of the Forehead and Temple. Facial Plastic Surgery Clinics of North America 13(2), 243–251 (2005)
8. Knize, D.: An anatomically based study of the mechanism of eyebrow ptosis. Plastic and Reconstructive Surgery 97(7), 1321–1333 (1996)
9. LaTrenta, G.S.: Atlas of Aesthetic Face & Neck Surgery. W.B. Saunders Company, Philadelphia (2004)
10. Lucas, B.D., Kanade, T.: An iterative image registration technique with an application to stereo vision. In: Proc 7th Intl. Joint Conf. on Artificial Intelligence (IJCAI), Vancouver, British Columbia, Canada, August 24-28, vol. 81, pp. 674–679 (1981)
11. Stuzin, J., Baker, T., Gordon, H.: The relationship of the superficial and deep facial fascias: Relevance to rhytidectomy and aging. Plastic and Reconstructive Surgery 89(3), 441–449 (1992)
12. Tang, J., Liu, J.: Ultrasonic measurement of scleral cross-sectional strains during elevations of intraocular pressure: Method validation and initial results in posterior porcine sclera. Journal of Biomechanical Engineering 134(9), 091007 (2012)

13. Vogt, M., Ermert, H.: Development and evaluation of a high-frequency ultrasound-based system for in vivo strain imaging of the skin. IEEE Transactions on Ultrasonics, Ferroelectrics, and Frequency Control 52(3), 375–385 (2005)
14. Warburton, M., Maddock, S.: Physically-based forehead animation including wrinkles. Computer Animation and Virtual Worlds (in press, 2014)
15. Weickenmeier, J., Itskov, M., Mazza, E., Jabareen, M.: A physically motivated constitutive model for 3D numerical simulation of skeletal muscles. International Journal for Numerical Methods in Biomedical Engineering 30(5), 545–562 (2014)
16. Weickenmeier, J., Jabareen, M.: Elastic-viscoplastic modeling of soft biological tissues using a mixed finite element formulation based on the relative deformation gradient. International Journal for Numerical Methods in Biomedical Engineering (in press, 2014)
17. Wu, T., Hung, A.L., Hunter, P., Mithraratne, K.: Modelling facial expressions: A framework for simulating nonlinear soft tissue deformations using embedded 3d muscles. Finite Elements in Analysis and Design 76, 63–70 (2013)
18. Wu, T., Mithraratne, K., Sagar, M., Hunter, P.J.: Characterizing facial tissue sliding using ultrasonography. In: Proceedings of the WCB, Singapore, pp. 1566–1569 (2010)

Author Index